ABOUT THE BOOK

Quantum chemistry is based on the postulates of Quantum Mechanics. The theory of quantum chemistry with an emphasis on their practical implications for the molecular modeller, and we will try to answer the questions of when to use a quantum chemical method instead of molecular mechanics, which quantum chemical method to choose, and what to expect from the quality of the results.

ABOUT THE AUTHOR

Phillips Wang, Born on 15 August 1978, he obtained his PhD. From University of York, Heslington, United Kingdom in 2010. His interests span from bio-resources fermentation technology to medical virology and nero science research. He is currently working as a research associate Professor at the University of Park Nottingham, United Kingdom.

Quantum Mechanics in Chemistry

Philips Wang

AMIGA PRESS INC.
DELHI (INDIA)

AMIGA PRESS INC.
H.O.: 28-A, Indra Park,
Chander Nagar, Delhi-110051 (India)
Tel.: 011-22504028, +91-98736-57000
E-mail: amigapress@gmail.com
amigaeditorial@gmail.com
amigapub@gmail.com
info@amigapress.co.in
Website : www.amigapress.co.in

This edition is published by Amiga Press Inc.

ISBN: 978-93-87295-33-9

First Published 2020

Printed and bound in India by Replika Press Pvt. Ltd.

Preface

Quantum chemistry is a branch of theoretical chemistry which applies quantum mechanics and quantum field theory to address problems in chemistry. One application of quantum chemistry is the electronic behavior of atoms and molecules relative to their chemical reactivity. Quantum chemistry lies on the border between chemistry and physics. Thus, significant contributions have been made by scientists from both fields. It has a strong and active overlap with the field of atomic physics and molecular physics, as well as physical chemistry.

Quantum chemistry is based on the postulates of Quantum Mechanics. The theory of quantum chemistry with an emphasis on their practical implications for the molecular modeller, and we will try to answer the questions of when to use a quantum chemical method instead of molecular mechanics, which quantum chemical method to choose, and what to expect from the quality of the results.

The quantum theory had its origin in an effort to account for the radiation by black bodies. A *black body* is generally understood to be a body which absorbs all the light falling upon it and, accordingly, can reflect none. Physicists, on the other hand, use the term in a much broader sense; by their interpretation a perfect black body is one which at all temperatures absorbs all the radiation falling upon it, whether it be visible light, ultraviolet, or infrared radiation.

It is highly desirable for students in all areas of chemistry to understand modern methods of electronic structure calculation, and this book has been written with this goal in mind.

– Author

Contents

❑❑❑

1.

Introduction

CRYSTAL STRUCTURE

A crystal structure is composed of a motif, a set of atoms arranged in a particular way, and a lattice. Motifs are located upon the points of a lattice, which is an array of points repeating periodically in three dimensions. The points can be thought of as forming identical tiny boxes, called unit cells, that fill the space of the lattice. The lengths of the edges of a unit cell and the angles between them are called the lattice parameters. The symmetry properties of the crystal are embodied in its space group. A crystal's structure and symmetry play a role in determining many of its properties, such as cleavage, electronic band structure, and optical properties.

CRYSTALLOGRAPHY

Amorphous solids are homogeneous and isotropic because there is no long range order or periodicity in their internal atomic arrangement. By contrast, the crystalline state is characterised by a regular arrangement of atoms over large distances. Crystals are therefore anisotropic – their properties vary with direction. For example, the interatomic spacing varies with orientation within the crystal, as does the elastic response to an applied stress. Engineering materials are usually aggregates of many crystals of varying sizes and shapes; these polycrystalline materials have properties which depend on the nature of the individual crystals, but also on aggregate properties such as the size and shape distributions of the crystals, and the orientation relationships between the individual crystals.

The randomness in the orientation of the crystals is a measure of texture, which has to be controlled in the manufacture of transformer steels, uranium fuel rods and beverage cans. The crystallography of interfaces connecting a d ja c e n t c rys ta ls candetermine the deformation behaviour of the polycrystalline aggregate; it can also influence the toughness through its effect on the degree of segregation of impurities to such interfaces.

MINERALS

Metals and ceramics are the basic materials of nuclear reactors. The regular arrangement of the atoms in such solids is termed a crystal lattice. Pure solids (whether elemental or binary) need not retain the same crystal structure from low temperature to the melting point; iron and uranium exhibit three crystal structures, each confined to a definite temperature interval, but uranium dioxide has only one. The change from one crystal structure to another is called a phase transformation. Such changes are the solid-solid equivalents of melting and vaporization. The crystal structure can exert a profound effect on material's behaviour in the environment of a nuclear reactor, as seen from the following examples:

- Void swelling can cause unacceptable dimensional changes in structures of fast reactors or fusion reactors (but not generally in light water reactors, except at higher doses and intermediate temperatures in austenitic stainless steels). The crystal structure of the austenitic form of steel is susceptible to this radiation effect but the different crystal structure of the ferritic form is immune until very high doses (>H"50 displacements per atom, dpa).
- Under fast-neutr on irradiation, Zircaloy cladding of LWRs undergoes growth in the axial direction. This effect is due to the anisotropy of the crystal structure of zirconium.
- The relation betw een c rystal a nisotropy and the directional dependence of macroscopic properties in zirconium was covered in Sect I.2.
- The utility of uranium metal as a nuclear fuel is diminished because of the volume change accompanying a solid phase transformation that occurs at relatively low temperature.Imperfections in the crystal lattice are even more important than the details of the atomic structure.
- Missing atoms in the atomic lattice called vacancies, or extra atoms in non-regular positions called interstitials, are responsible for the mobility of atoms in the solid
- Many of the mechanical properties of metals are governed by a linear defect in the atomic structure called a dislocation
- All structural alloys and ceramic fuels consist of a large number of small crystals, or grains, joined at their peripheries. The discontinuities in the crystal structure at these internal surfaces are called grain boundaries. Their presence influences a variety of mechanical and transport properties, including cracking in steel and fission gas movement in fuel.

For elemental solids, only 14 distinct structures, or spatial arrangements of atoms, are possible. Each crystal structure is characterized by a unit cell, which is a small group of atoms that contains the unique features of the particular lattice type. A macroscopic crystal is composed of large numbers of unit cells stacked together.

The elemental and binary solids important in LWRs are restricted to only a few of the 14 possible lattice types, mainly those with the highest degrees of symmetry and the tightest packing of atoms. With one exception, these unit cells are parallelepipeds (all angles 90o) with side lengths denoted by a, b, and c. The exception is a structure whose unit cell is a hexagonal prism with only two characteristic dimensions. Only the parallelepiped with all sides of equal length (*i.e.*, a cube) is isotropic. All the others are anisotropic, meaning that properties differ in the principal coordinate directions.

LATTICE

We will describe structures of crystalline solids in terms of the lattice. A lattice is a regular, repeating pattern of points in space. Wallpaper patterns are examples of 2-dimensional lattices. A wallpaper pattern and the equivalent lattice. The lattice is more fundamental than the particular wallpaper pattern shown, because wallpaper patterns using many different design units could be generated from the same lattice.

The key feature of the lattice is the arrangement of points. A small collection of lattice points that contains all of the spatial information in the lattice pattern is called a unit cell. The unit cell represents the lattice, because the whole lattice can be reproduced from the unit cell by moving (translating) the unit cell in directions parallel to its sides by distances equal to the lengths of its sides.

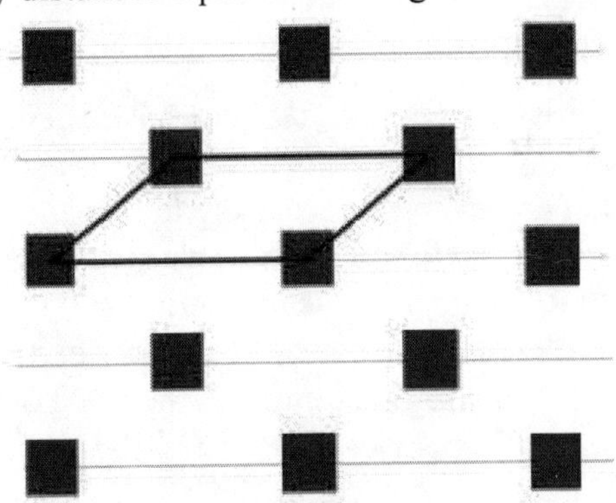

Fig. Unit Cell in the Lattice

Movement of this collection in directions parallel to its sides does indeed reproduce the entire lattice. However, the unit cell in Figure is simpler than that in Figure because it involves fewer points—one point at each of its 4 corners. A unit cell having lattice points at its corners only is called a primitive unit cell. The unit cell in Figure is a non-primitive unit cell because it has an additional point at its center. If it is possible to represent a lattice with a primitive unit cell, we will do so. Crystalline solids are represented by 3- dimensional lattices, one example of which is the simple cubic lattice. This is a primitive unit cell because it has points only at the corners. Movement of the unit cell by the distance, d, in directions parallel to its 3 dimensions reproduces the entire simple cubic lattice. The simple cubic lattice is analogous to an orderly arrangement of building blocks stacked face to face in 3-dimensions.

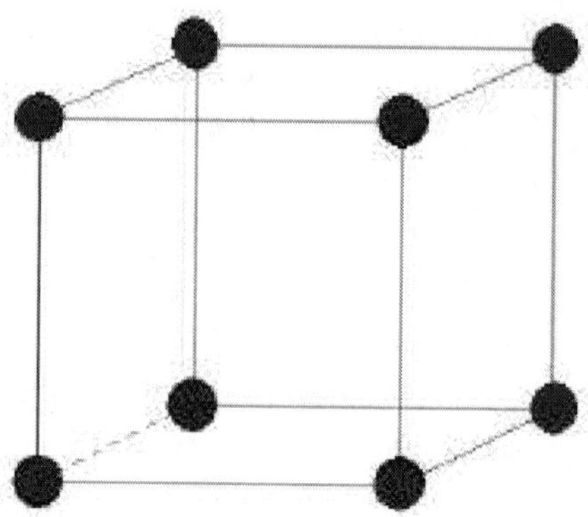

Fig. cubic lattice

It was shown in the 1800s by Bravais that there are only 7 shapes (called crystal types) that a unit cell can have and only 14 possible distributions of points (lattice types) within these 7 shapes. The 7 shapes, characterized by the relative magnitudes of the repeat distances in three directions and the angles between these directions.

CLOSE PACKING

What is the most efficient way of arranging spheres in 3-dimensional space? We take "efficient" to mean economical in terms of space. We begin by building one layer of spheres that are arranged as closely to one another as possible. Note that this arrangement is identical to that pictured earlier in Figure. Spheres are lined up in straight rows, with the spheres in one row fitting sn ugly into the inden tations in adjacent rows. Each sphere is surrounded by a perfect hexagon of spheres, all touching it. There are numerous gaps, or holes, in the layer, of triangular shape. For descriptive purposes here, we label these holes "u" if the triangle has a single vertex pointing up, or "d" if the triangle has vertex down.

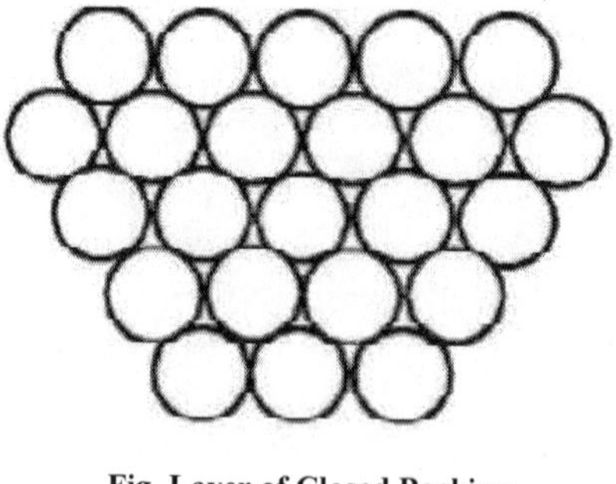

Fig. Layer of Closed Packing

We now add a single sphere of a second layer on top of the first. The added sphere rests in a low point of the first layer, corresponding with a triangular hole, in contact with 3 spheres of the first layer. The 4 touching spheres lie at the vertices of a regular tetrahedron. There remains some unfilled space at the center of this 4-sphere group that is called a tetrahedral hole. Each time we add a sphere to the second layer, we create a tetrahedral hole between the layers. When additional spheres are added to the second layer, it turns out that they must all go in either "u" holes or "d" holes, but not both. Thus only half of the triangular depressions in the first layer can be used for placement of 2nd-layer spheres. Triangular holes in the 2nd layer lie directly over the unused holes of the first layer, so that channels pass right through both layers.

The point at the center of such a channel is surrounded by 6 spheres arranged as 2 equilateral triangles, one triangle in the 1st and one in the 2nd layer. These 6 spheres form a regular octahedron, a geometrical solid with 6 vertices and 8 equilateral triangular faces. The space surrounded by the 6 spheres is called an octahedral hole. An octahedral hole is larger than a tetrahedral hole because 6 spheres cannot be crowded as closely together as 4 spheres. There are 2 ways in which a third layer of spheres can be added to the existing 2 layers in close-packed fashion. In the first way, spheres are place directly over spheres of the first layer. The third and first layers then have spheres in the same locations. The 4th layer then goes over the second, the 5th over the 3rd, and so on, so a pattern that repeats every other layer is obtained.

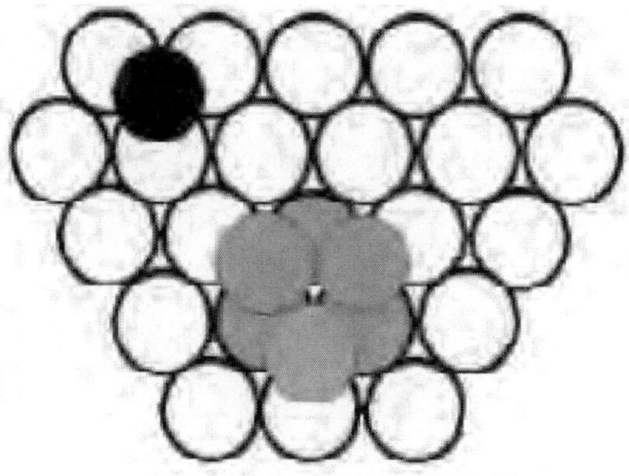

Fig. Black-shaded Sphere

The unit cell is hexagonal in shape; the 1212... close-packing arrangement is therefore called hexagonal close packing (hcp). Note that the coordination number (the number of nearest neighbours) of any sphere in hcp is 12.

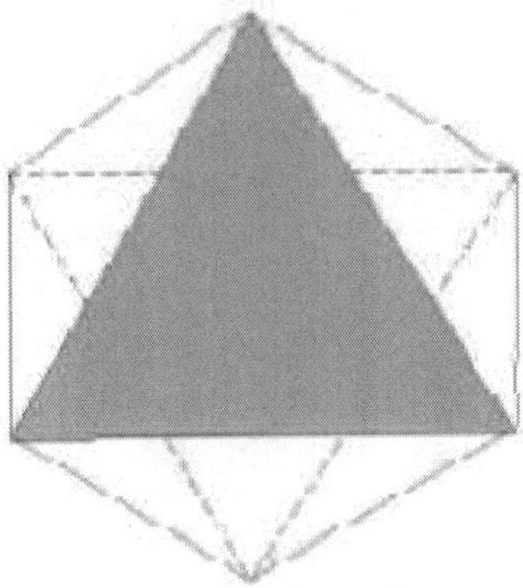

Fig. Regular Octahedron

The second way of adding a third layer is to place spheres directly over the octahedral holes between the first two layers. In this case, the third-layer spheres do not lie over spheres in layer 1. This is still a close-packing arrangement because the coordination number of any sphere is 12, just as in hcp. The 4th layer of spheres is then placed directly over the spheres in the first layer.

This type of close packing is designated 123123... to indicate that the pattern repeats after 3 layers. Although it may not be obvious from the figure, this unit cell is a cube (this is a situation in which a 2-dimensional drawing does not adequately convey what is true). This type of close packing is therefore called cubic close packing (ccp). Hcp and ccp are the two ways of close packing spheres in 3 dimensions. They have several features in common:

Spheres occupy 74 per cent of the available space; Each sphere has 12 nearest neighbours that touch it (coordination number); The number of octahedral holes in the lattice is exactly the same as the number of spheres in the lattice, and there are twice as many tetrahedral holes as spheres! This is not obvious and will be addressed again later. There are a number of substances that crystallize in one or the other of these two packing arrangements. Beryllium, magnesium, and zinc metal all form hexagonal close packed lattices. Nickel, copper, gold, and aluminum form ccp lattices. Cubic close packing is particularly common. We will take a closer look at cubic lattices now.

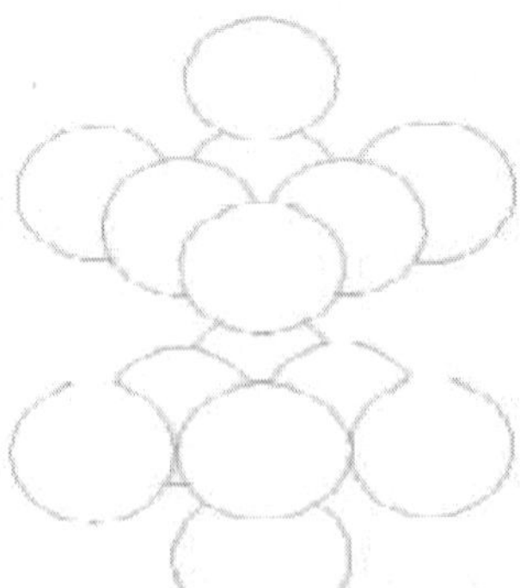

Fig. Unit Cell for 123... Close Packing

The three Bravais cubic lattices are shown in Figure. The simple cubic (sc) lattice is primitive, with points (spheres or atoms) only at the vertices of the cube. Metallic polonium and non-metallic dioxygen (O_2) crystallize with the simple cubic lattice. The coordination number of an atom in the simple cubic lattice is 6.

To determine this, pick one atom in the unit cell; find an atom that appears to be closest to the atom you picked; then count the number of other atoms that are at this same distance from the atom you picked. In counting coordination numbers, you may have to go outside of the single unit cell pictured. In the simple cubic lattice, identical unit cells surround the pictured one on all sides. Some of the nearest neighbours of your picked atom lie in these neighbouring unit cells. Finally, the simple cubic unit cell contains only

1 atom, despite the fact that you see 8 atoms drawn in the figure. We say that the simple cube has 1 atom per unit cell (1 atom/uc).

The procedure for counting atoms in unit cells is described in the paragraph below. The body-centered cubic (bcc) lattice is non-primitive, with points at the corners and the center of the unit cell (the center of a cube is called the cube body center). Metallic iron and barium crystallize in the bcc lattice. The coordination number is 8 for this lattice (count the nearest neighbours of the body-center atom), and the body-centered cube has 2 atoms per unit cell. Finally, the face-centered cubic (fcc) lattice is also non-primitive with points at each corner and in the center of each of the 6 faces of the cube. Metallic nickel and copper adopt this structure. The coordination number is 12 for this lattice, which means that it must be close packed! The face-centered cubic lattice and the cubic close packed lattice are identical. It is very difficult to see this in two-dimensional drawings. To see it, you must imagine looking along a body diagonal of the face-centered cube. A body diagonal runs on a straight line from one corner, through the body center of the cube, to the diagonally opposite corner.

As you look along the body diagonal a single sphere; a triangle consisting of 6 spheres; a second triangle of 6 spheres, but rotated 60° with respect to the first one; and finally another single sphere immediately below the first one. The cube is admittedly difficult to visualize from the drawings. The face- centered cubic lattice has 4 atoms per unit cell.

We now address the problem of determining the number of atoms per unit cell. This is quite simple once it is realized that atoms are generally shared among several unit cells and therefore must be partitioned accordingly. An atom at a corner (vertex) of a cube is shared among 8 unit cells. To see this, imagine the simple cube to be a building block, and focus on the upper right front corner of the building block. Imagine that this corner is labelled "a". In your mind, place identical blocks on top, to the right, and in front of the block in the figure.

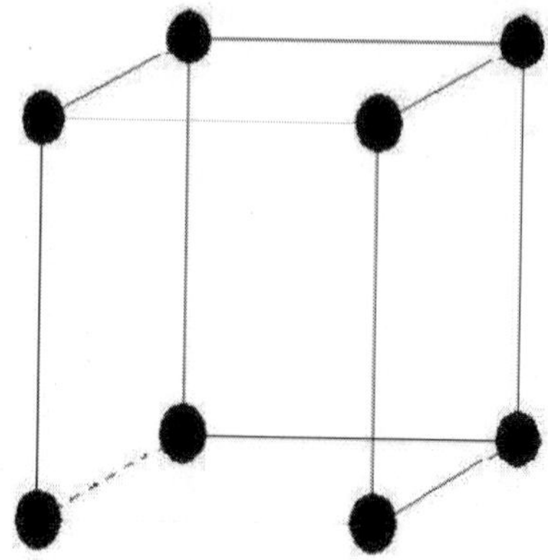

Fig. Three Bravais Cubic Lattices

Then place additional blocks in the remaining places adjoining these blocks to generate the 8-block arrangement. Point "a" is now at the center of this collection, and is simultaneously a corner for each of the 8 blocks; 8 blocks share a vertex.

This point cannot be said to belong entirely to any particular block. Instead, each block gets 1/8 of it. The counting rule for a corner atom, then, is that only 1/8 of it is assigned to a single unit cell.

Similar thinking leads to the conclusion that a face-center atom is shared between 2 unit cells, so 1/2 of it is assigned to a particular unit cell; and a body-center atom is not shared at all, so it belongs entirely to the unit cell that it is found in. The atom count for a body-centered unit cell then goes like this: atoms/*uc* = 8 corner atoms × 1/8 + 1 body-center atom × 1.

We will have occasion also to partition points located along edges of the cube. Edges are the lines connecting adjacent vertices. Since an edge is common to 4 cubes arranged in a square, 1/4 of any point along an edge is assigned to a particular cube. The point-counting rules are summarized in.

Table. Atom Counting Rules for Cubic Unit Cells

Atom location	Fraction Assigned to uc
Corner	1/8
Edge	1/4
Face	1/2
Body-center	1

If, as we have said, the face-centered cubic lattice is in fact cubic close packed, then there should be tetrahedral and octahedral holes in the fcc lattice. Further, we should find 4 octahedral and 8 tetrahedral holes in a fcc unit cell based on the the last-listed common feature of close-packed lattices above. To locate these holes it is convenient to place our cube in a coordinate system. The origin of coordinates is located at the lower left front corner of the cube, with the x-axis along the bottom front edge, the y-axis along the left vertical edge, and the z axis along the left bottom edge.

We specify distances along the 3 axes in terms of fractions of the cube edge length. The set of coordinates (1,0,0) means to proceed 1 edge length along x. These are the coordinates of the lower right front vertex of the cube. Similarly, (1/2,1/2,1/2) specifies the body center of the cube, labelled 2 in the figure. Now examine the face-centered cube.

We find the most obvious octahedral hole at the body center. This point is surrounded by 6 atoms all at the same distance from it. These are the atoms at the face centers of the cube. Similar octahedral holes are located at the cube edge centers. For example, the point labelled 3 is an octahedral hole. It is surrounded at equal distance by the atoms at (1/2,1/2,1), (1,0,1), (1,1,1), and (1,1/2,1/2).

The remaining 2 atoms are not part of the unit cell shown, but can still be located with coordinate sets (1 1/2,1/2,1) and (1,1/2,-1 1/2). (Translating these coordinates is good practice in the use of the unit cell coordinate system.) The cube has 12 edges, so there are 12 holes of this type. The counting rules tell us that only 1/4 of each of these edge-centered octahedral holes belongs to the unit cell of interest. The total number of octahedral holes in the unit cell is then 1(body center) + 12(1/4) = 4. The number of octahedral holes is the same as the number of atoms in the unit cell.

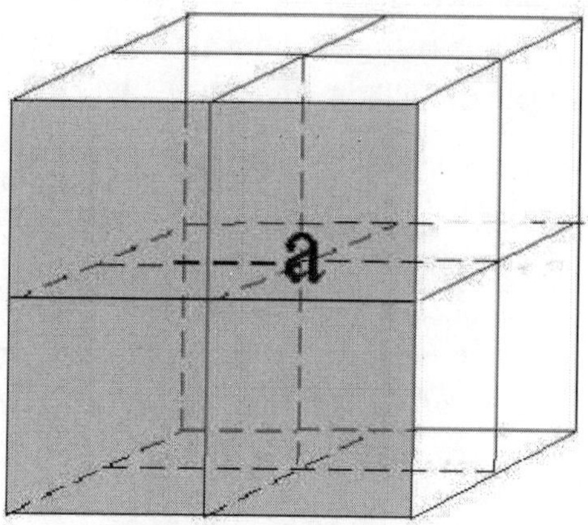

Fig. **Blocks to Generate the 8-block**

It is defined by the four shaded atoms, and has coordinates (1/4,1/4,1/4). There are 7 other tetrahedral holes in the unit cell, all with coordinate combinations involving 1/4 and 3/4, and all defined by 1 corner and 3 face-center atoms. Try to locate them in the figure, and write down their coordinates.

Each tetrahedral hole lies entirely within the cube body, so none of them are shared with other unit cells. The number of tetrahedral holes per fcc unit cell is therefore 8, twice the number of atoms in the unit cell. Octahedral and tetrahedral holes are important in the discussion of ionic compounds.

Octahedral and tetrahedral holes are important in the discussion of ionic compounds.

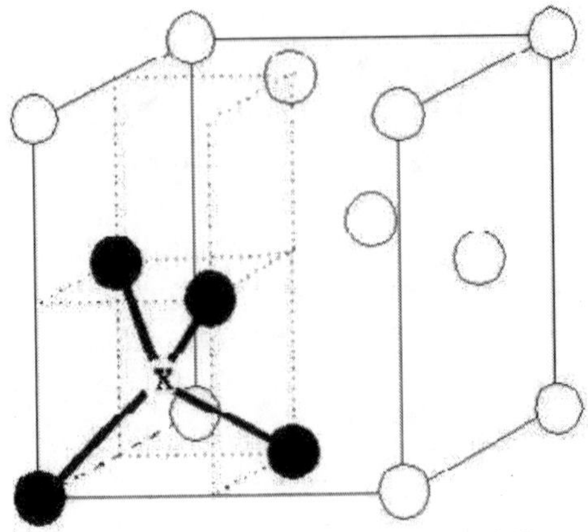

Fig. Tetrahedral Hole in the Fcc Unit Cell

BRAVAIS LATTICES

The number of ways in which points can be arranged regularly in three dimensions, such that the stacking of unit cells fills space, is not limitless; Bravais showed in 1848 that all possible arrangements can be represented by just fourteen lattices. The fourteen Bravais lattices can be categorised into seven crystal systems (cubic, tetragonal, orthorhombic, trigonal, hex agonal, monoclinic and triclinic); the cubic system contains for example, the cubic-P, cubic-F and cubic-I lattices. Each crystal system can be characterised uniquely by a set of defining symmetry elements, which any crystal within that system must possess as a minimum requirement.

Table. The Crystal Systems

System	Conventional unit cell	Defining symmetry
Triclinic	$a_1 \neq a_2 \neq a_3, a \neq \beta \neq \gamma$	monad
Monoclinic	$a_1 \neq a_2 \neq a_3, \alpha = \beta, \beta > 90^o$	1 diad
Orthorhombic	$a_1 \neq a_2 \neq a_3, \alpha = \beta, \beta > 90^o$	3 diad
Tetragonal	$a_1 \neq a_2 \neq a_3, \alpha = \beta, \beta > 90^o$	1 tetrad
Trigonal	$a_1 - a_2 \neq a_3, \alpha - \beta - \beta - 90^o$	1 triad
Hexagonal	$a_1 = a_2 \neq a_3, \alpha = \beta = 90^o, \beta = 120^o$	1 hexad
Cubic	$a_1 = a_2 = a_3, \alpha = \beta = \gamma = 90^o$	4 triads

The Bravais lattices are each as a projection along the a_3 axis. Projections like these are useful as simple representations of three–dimensional objects. The coordinate of any point with respect to the a_3 axis is represented as a fraction of a_3 along the point of interest; points located at $0a_3$ are unlabelled; translational symmetry requires that for each lattice point located at $0a_3$, there must exist another at $1a_3$.

LINEAR COMBINATIONS

Any vector u can be represented as a linear combination of the basis vectors ai of the unit cell (i = 1, 2, 3):

$$u = u_1a_1 + u_2a_2 + u_3a_3$$

and the scalar quantities $u1$, u_2 and u_3 are the components of the vector u with respect to the basis vectors a_1, a_2 and a_3. Once the unit cell is defined, any direction u within the lattice can be identified uniquely by its components [u_1 u_2 u_3], and the components are called the Miller indices of that direction and are by convention enclosed in square brackets.

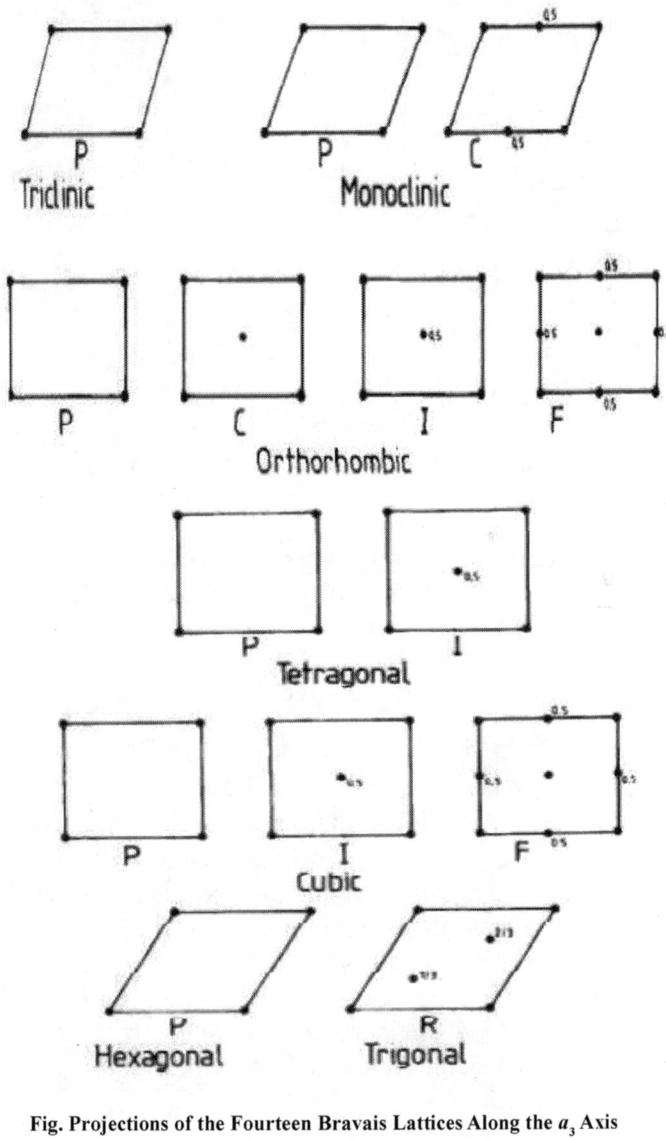

Fig. Projections of the Fourteen Bravais Lattices Along the a_3 Axis

It is sometimes the case that the properties along two or more different directions are identical. These directions are said to be equivalent and the crystal is said to possess symmetry.

For example, the [1 0 0] direction for a cubic lattice is equivalent to the [01 0], [0 0 1], [0 1 0], [0 0 1] and [1 0 0] directions; the bar on top of the number implies that the index is negative.

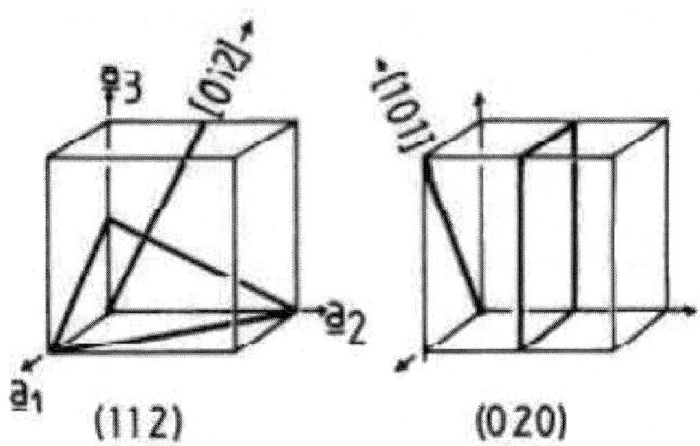

Fig. Miller Indices for Directions and Planes

The indices of directions of the same form are conventionally enclosed in special brackets, *e.g.* < 1 0 0 >. The number of equivalent directions within the form is called the multiplicity of that direction, which in this case is 6.

PLANES

If a plane intersects the a1, a2 and a3 axes at distances x1, x2 and x3 respectively, relative to the origin, then the Miller indices of that plane are given by (h1 h2 h3) where:

$$h_1 = \phi\frac{a_1}{x_1},\ h_2 = \phi\frac{a_2}{x_2},\ h_3 = \phi\frac{a_3}{x_3}$$

ϕ scalar which clears the numbers *his* off fractions or common factors. Note that *xi* are negative when measured in the “*ai* directions.

The intercept of the plane with an axis may occur at 1, in which case the plane is parallel to that axis and the corresponding Miller index will be zero. Miller indices for planes are by convention written using round brackets: (h1 h2 h3) with braces being used to indicate planes of the same form: {h1 h2 h3}.

The Reciprocal Lattice

The reciprocal lattice is a special co–ordinate system. For a lattice represented by basis vectors a1, a2 and a3, the corresponding reciprocal basis vectors are written a_1^*, a_2^* and a_3^*, such that:

$$a_1^* = \frac{(a_2 \wedge a_3)}{(a_1 . a_2 \wedge a_3)}$$

$$a_2^* = \frac{(a_3 \wedge a_1)}{(q_1 . q_2 \wedge q_3)}$$

$$a_3^* = \frac{(a_1 \wedge a_2)}{a_1 . a_2 \wedge a_3}$$

In equation 2a, the term $(a_1 . a_2 \wedge a_3)$ represents the volume of the unit cell formed by ai, while the magnitude of the vector $(a_2 \wedge a3)$ represents the area of the (100)A plane. Since $(a_2 \wedge a_3)$ points along the normal to the (100)A plane, it follows that a^* also points along the normal to (100) A and that its magnitude || is the reciprocal of the spacing of the (100)A planes.

The components of any vector referred to the reciprocal basis represent the Miller indices of a plane whose normal is along that vector, with the spacing of the plane given by the inverse of the magnitude of that vector. For example, the reciprocal lattice vector u* = (123) is normal to planes with Miller indices (123) and interplanar spacing 1/|u*|.

$a_i a_j^* = 1$ *when* $i = j$, *and* $ai.a_j^* = 0$ *when* $i = / j$ or in other words $a_i a_j^* = \delta_{ij}$

δ_{ij} is the Kronecker delta, which has a value of unity when $i = j$ and is zero when $i = /j$

ELECTRONS IN METALS

Metals are electrical conductors because some of the electrons are able to move freely, even though they move in a periodic array of positive metal ions. Electrons are waves, characterised by a wave number k

$$k = \pm \frac{2\pi}{\lambda}$$

where λ is the wavelength. The lattice does not affect the motion of these electrons except at critical values of k, where the lattice planes reflect the el ectrons†. For a square latti ce, these critical val ues of λ are given by

$$k = \frac{n\pi}{a \sin\theta}$$

where a is the lattice parameter and μ is the direction of motion. The component of k along the x axis is $kx = k \sin \mu$ and since $\mu = 0$, it follows that reflection occurs when, and similarly when,

$$kx = \pm \frac{1/4}{a}$$

and similarly when,

$$ky = \pm \frac{1/4}{a}.$$

This can be represented in k–space, which is reciprocal space with a reciprocal lattice parameter of magnitude 2 π /a. In Fig. below, electrons are reflected in the lattice whenever k falls on any point on the square ABCD. This square is known as the first Brillouin zone boundary. Electron energy contours are frequentl plotted in k–space. Inside a Brillouin zone, the energy of electrons can increase 'smoothly' with k; there is then a discontinuity (an energy gap) at the zone boundary. This is followed by a second Brillouin zone ($n = 2$ in equation 3) and so on.

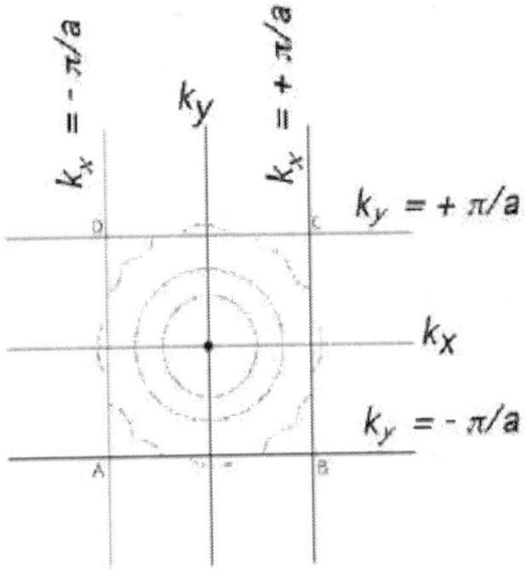

Fig. The Shape of the First Brillouin Zone for the Face–centred Cubic Lattice, Drawn in *k*–space.

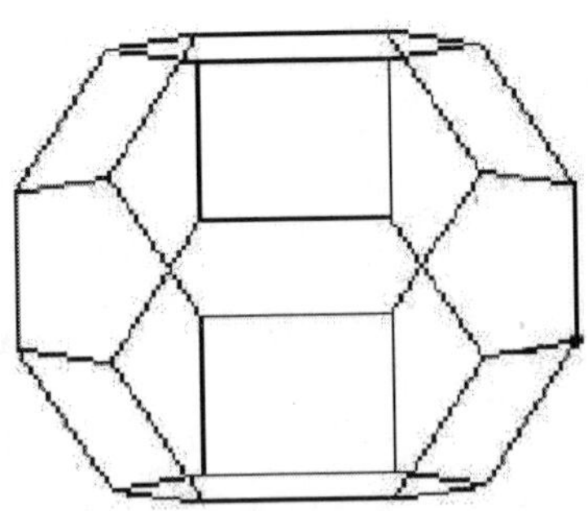

Fig. The Shape of the First Brillouin Zone for a Three–Dimensional Face Centred–cubic Lattice

UNIT CELL

The crystal structure of a material or the arrangement of atoms in a crystal structure can be described in terms of its unit cell. The unit cell is a tiny box containing one or more motifs, a spatial arrangement of atoms. The units cells stacked in three-dimensional space describes the bulk arrangement of atoms of the crystal. The unit cell is given by its lattice parameters, the length of the cell edges and the angles between them, while the positions of the atoms inside the unit cell are described by the set of atomic positions (*xi*, *yi*, *zi*) measured from a lattice point.

Although there are an infinite number of ways to specify a unit cell, for each crystal structure there is a conventional unit cell, which is chosen to display the full symmetry of the crystal. However, the conventional unit cell is not always the smallest possible choice. A primitive unit cell of a particular crystal structure is the smallest possible volume one can construct with the arrangement of atoms in the crystal such that, when stacked, completely fills the space. This primitive unit cell will not always display all the symmetries inherent in the crystal. A Wigner-Seitz cell is a particular kind of primitive cell which has the same symmetry as the lattice. In a unit cell each atom has an identical environment when stacked in 3 dimensional space. In a primitive cell, each atom may not have the same environment.

SYMMETRY

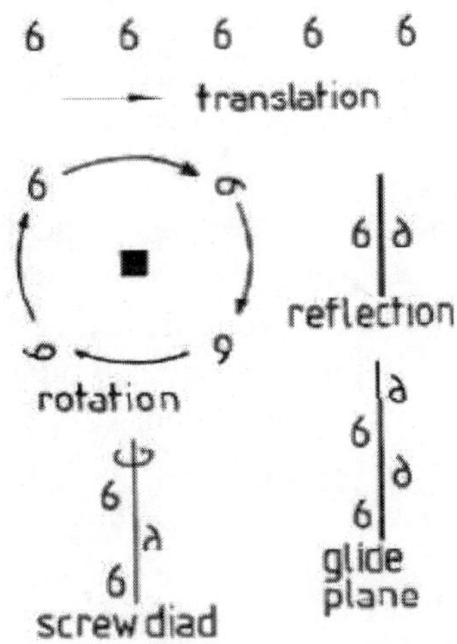

Fig. An Illustration of some Symmetry Operations

Although the properties of a crystal can be anisotropic, there may be different directions along which they are identical. These directions are said to be equivalent and the crystal is said to possess symmetry. That a particular edge of a cube cannot be distinguished from any other is a measure of its symmetry; an orthorhombic parallelepiped has lower symmetry, since its edges can be distinguished by length. Some symmetry operations are illustrated in ess ence, th ey tra nsform a spatial arra ngem ent into anoth er w hich is indistinguishable from the original.

The rotation of a cubic lattice through 90° about an axis along the edge of the unit cell is an example of a symmetry operation, since the lattice points of the final and original lattice coincide in space and cannot consequently be distinguished. We have already encountered translational symmetry when defining the lattice; since the environment of each lattice point is identical, translation between lattice points has the effect of shifting the origin.

Symmetry Operations

An object possesses an n-fold axis of rotational symmetry if it coincides with itself upon rotation about the axis through an angle $360°\pm/n$. The possible angles of rotation, which are consistent with the translational symmetry of the lattice, are 360°, 180°, 120°, 90°and 60° for values of n equal to 1, 2, 3, 4 and 6 respectively. A five–fold axis of rotation does not preserve the translational symmetry of the lattice and is forbidden. A one-fold axis of rotation is called a monad and the terms diad, triad, tetrad and hexad correspond to $n = 2, 3, 4$ and 6 respectively.

All of the Bravais lattices have a centre of symmetry. An observer at the centre of symmetry sees no difference in arrangement between the directions $[u_1\ u_2\ u_3]$ and $[-u_1\ -u_2\ -u_3]$. The centre of symmetry is such that inversion through that point produces an identical arrangement but in the opposite sense.

A rotoinversion axis of symmetry rotates a point through a specified angle and then inverts it through the centre of symmetry such that the arrangements before and after this combined operation are in coincidence. For example, a three-fold inversion axis involves a rotation through 120° combined with an inversion, the axis being labelled $\overline{3}$.

A rotation operation can also be combined with a translation parallel to that axis to generate a screw axis of symmetry. The magnitude of the translation is a fraction of the lattice repeat distance along the axis concerned. A 31 screw axis would rotate a point through 120° and translate it through a distance t/3, where t is the magnitude of the shortest lattice vector along the axis. A 32 operation involves a rotation through 120° followed by a translation through 2t/3 along the axis. For a right–handed screw axis, the sense of rotation is anticlockwise when the translation is along the positive direction of the axis.

A plane of mirror symmetry implies arrangements which are mirror images. Our left and right hands are mirror images. The operation of a 2 axis produces a result which is equivalent to a reflection through a mirror plane normal to that axis. The operation of a glide plane combines a reflection with a translation parallel to the plane, through a distance which is half the lattice repeat in the direction concerned. The translation may be parallel to a unit cell edge, in which case the glide is axial; the term diagonal glide refers to translation along a face or body diagonal of the unit cell. In the latter case, the translation is through a distance which is half the length of the diagonal concerned, except for diamond glide, where it is a quarter of the diagonal length.

CLASSIFICATION OF CRYSTALS BY SYMMETRY

The defining property of a crystal is its inherent symmetry, by which we mean that under certain operations the crystal remains unchanged. For example, rotating the crystal 180 degrees about a certain axis may result in an atomic configuration which is identical to the original configuration. The crystal is then said to have a twofold rotational symmetry about this axis. In addition to rotational symmetries like this, a crystal may have symmetries in the form of mirror planes and translational symmetries, and also the so-called compound symmetries which are a combination of translation and rotation/ mirror symmetries. A full classification of a crystal is achieved when all of these inherent symmetries of the crystal are identified.

IONIC COMPOUNDS

The ideas discussed thus far apply to crystals in which the units—spheres or atoms—are all identical. Crystals of pure metals, with close-packed metal atoms, and even crystals of molecular solids (for example, methane, CH_4, in which there is a methane molecule at each lattice point) are examples of real situations in which these ideas apply. Ionic compounds are different because there are two distinctly different units—the cation and the anion—that must be accommodated in the lattice. The resulting lattices are often called compound lattices because they contain two different components. The cation and the anion differ in charge, of course, and in chemical identity (usually).

They also differ in size. Generally speaking, cations are smaller than anions, for easily-understood reasons. Cations contain an excess of protons over electrons. The Coulomb pull of the excess positive charge causes the electron cloud to contract, so that it is smaller than in the corresponding atom (*e.g.*, Na^+ is smaller than Na). Anions contain an excess of electrons over protons. The repulsion between the electrons exceeds that in the corresponding neutral atom, causing an expansion of the electron cloud (*e.g.*, Cl- is larger than Cl). Except for a few rare cases, the cation in an ionic compound is smaller than the anion. In forming a lattice, the space requirements are more important for the larger anion, which tends to dictate which lattice is adopted.

The (smaller) cations then fit into the holes, or interstices, of the anion lattice. Many common ionic compounds have lattices that consist of cubic close packed a nions, w ith ca tion s in tetra hedral and/ or oc tah edra l h oles. Stoichiometries of up to 3 cations per anion can be accommodated in this framework. We now discuss some common ionic lattices.

SODIUM CHLORIDE LATTICE

It can be described as a cubic close packed (face-centered cubic) anion lattice, with Na^+ ions in the octahedral holes. The coordination number (number of nearest neighbours of opposite charge) of each Na^+ ion is 6, and of each Cl^- ion is 6. The cation and anion must have equal coordination numbers because the stoichiometry of the compound is 1:1 (1 cation per anion).

The unit cell contains 4 formula units of NaCl (4 Cl- ions, which define the fcc unit cell; 4 Na^+ ions because there are 4 octahedral holes per unit cell). Many ionic compounds exhibit the NaCl lattice, including most of the group 1 halides, most of the group 2 oxides, and a number of d-metal oxides.

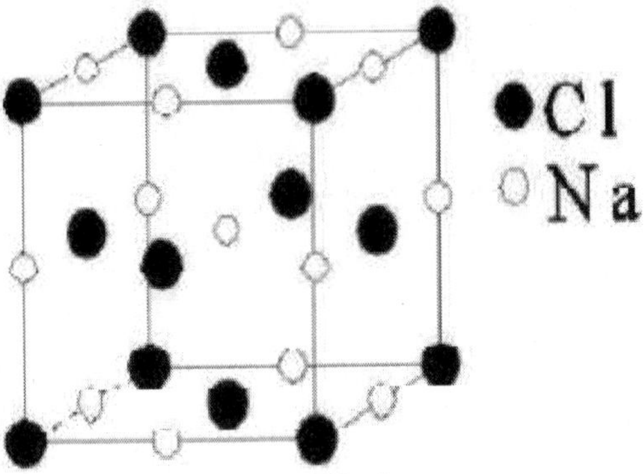

Fig. NaCl lattice

The Zinc Blende Lattice. Zinc blende is the common name for one crystal modification of the common mineral, ZnS. The other form is known as wurtzite. Wurtzite consists of an hcp array of S^{2-} ions with Zn^{2+} ions in one-half of the tetrahedral holes; it will not concern us further here, because we are restricting our attention to cubic lattices.

Zinc blende consists of a ccp array of S^{2-}, with Zn^{2+} in half of the tetrahedral holes. Since there are 8 tetrahedral holes in the sulfide ion lattice, but only 4 S^{2-} ions, only half of the holes are used to accommodate 4 Zn^{2+} ions. This is consistent with the 1:1 stoichiometry of the compound. The zinc ions occupy tetrahedral holes so that they can be as far apart from one another as possible to minimize repulsions between positive ions.

If the first Zn^{2+} ion occupies the hole at (1/4,3/4,1/4), the other three will be in holes at (3/4,1/4,1/4), (1/4,1/4,3/4), and (3/4,3/4,3/4). You should locate these holes and verify that the Zn^{2+} ions describe a tetrahedron. The cation and anion coordination numbers are both 4 in zincblende. As for NaCl, there are 4 formula units per unit cell. Other ionic compounds that adopt the zinc blende lattice include BeS, CdS, HgS, and AgI.

CALCIUM FLUORIDE LATTICE

This compound has formula CaF2, and exhibits the lattice. The Ca^{2+} ion is virtually the same size as the F^- ion, one of those rare situations referred to above, and forms a face-centered cubic lattice. The F- ions fill all of the tetrahedral holes in the cation lattice. Since there are 8 such holes per 4 Ca^{2+} ions, the stoichiometry is nicely accommodated. The coordination number of the Ca^{2+} ion is 8, and that of the F^- ion is 4. There are again 4 formula units per unit cell. The fluorite structure is very common for ionic compounds of 1:2 (or 2:1) stoichiometry.

The Cryolite Lattice. Cryolite is the common name for the mineral, Na_3AlF_6. Though rare, this substance is of extreme commercial value, because it is used in the process of extraction of aluminum from bauxite ore. Cryolite consists of Na^+ cations and AlF_6^{3-} anions. It adopts a lattice in which the anions are cubic close packed (fcc) and the cations occupy all of the octahedral and tetrahedral holes.

The Diamond Lattice. Diamond is one crystal modification of the element carbon. It is, of course, not ionic, nor even a compound. It is dealt with here because it, too, is based on a fcc unit cell. The diamond lattice is just like the zinc blende lattice, with carbon atoms in place of both Zn^{2+} and S^{2-} ions. We therefore describe it as a fcc array of carbon atoms with carbon atoms in one- half the tetrahedral holes. The coordination number of each carbon atom in diamond is 4.

CESIUM CHLORIDE LATTICE

It is a more open lattice than those considered thus far. This lattice consists of a simple cubic array of chloride ions, with Cs^+ ions in cubic holes. A cubic hole is the space at the body center of a simple cubic unit cell. There is 1 formula unit of CsCl per unit cell, and the coordination number of both cation and anion is 8. Ionic compounds that adopt the CsCl lattice include CsBr, CsI, the halides of thallium(I), and NH_4Cl.

RADIUS RATIO RULES

Compounds of 1:1 stoichiometry commonly adopt one of three lattices discussed above: the zinc blende lattice, the rock salt lattice, or the cesium chloride lattice. Of interest now is whether or not we can predict, or at least understand, why a particular compound favours one lattice over the others. A number of factors are important in choice of lattice, but one of these dominates in importance: the relative sizes of cation and anion.

Because we view the cation as occupying holes in the anion lattice, we begin by calculating the sizes of the various hole types in relation to the size of the anion involved. The most straightforward calculation is for the octahedral hole.The figure shows one face of the fcc unit cell, with the anions just touching along a face diagonal of the cube. There is an octahedral hole at the center of each of the 4 cube edges shown; we focus on the right edge.

The size of the octahedral hole will be taken as the distance from the edge center to, say, the top right anion, along the cube edge. (The distances from the hole to the bottom right anion and to the face-centered anion are the same.) We can express the right edge length as a sum of anion and hole radii as in equation:

$$a = 2r- + 2r+$$

Here r- is the anion radius, r+ the hole radius. (The + designation is used in anticipation of putting a cation in the hole.) Similarly, we express the length of the face diagonal (fd) in terms of anion radii.

This is easy, since the anions touch along this line:

$$fd = 4r-$$

Finally, we relate the fd and edge length via the Pythagorean Theorem: $(fd)2 = 2a2$

Take the square root of both sides of 7-4-3:

$$fd = a(2)1/2$$

Substitute 7-4-4 in 7-4-2 and rearrange for a:

$$a = 2(2)1/2\ r-$$

Substitute for a in 7-4-1 and solve for r+:

$$r+ = ((2)1/2-1)\ r- = 0.414\ r-$$

The octahedral hole is somewhat less than half as large as the anions that create the hole. By similar geometric methods, equations 7-4-7 and 7-4-8, for tetrahedral and cubic holes respectively, can be developed:

$$r+ = 0.224\ r-$$

$$r+ = 0.732\ r-$$

You should try to obtain these expressions, making use of right triangles and Pythagorean relationships. (In addition to equation which relates the face diagonal and edge length of a cube, equation relates the cube body diagonal (bd) to the edge length:

$$bd = (3)1/2a$$

According to Equations 7-4-6 to 7-4-8, a tetrahedral hole is about 1/2 the size of an octahedral hole and about 1/3 the size of a cubic hole. Only cations that are relatively small compared with the anion can be accommodated in tetrahedral holes. As cation size increases relative to anion size, the cation will prefer a larger hole, with a larger coordination number of anions. Equations 7-4-6 to 7-4-8 are the basis for the following radius-ratio rules.

If the ratio of the cation and anion radii (r+/r–) lies in the range between 0.224 and 0.414, the cation prefers to occupy a tetrahedral hole. If *r+/r–* lies in the range between 0.414 and 0.732, the cation prefers an octahedral hole. If *r+/r–* exceeds 0.732, the cation prefers a cubic hole.

These rules make good sense. If *r+/r–* < 0.224, the cation is too small for the tetrahedral hole. The anions will be touching, but will not be touching cations. Since it is the cation-anion attractions that stabilize the lattice, whereas anion-anion or cation-cation repulsions tend to destabilize it, this is not a favourable situation. When *r+/r–* = 0.224, the cation just fits the tetrahedral hole. Although anions are still touching, they now touch cations as well. This is a stable arrangement. As *r+/r–* becomes > 0.224, the cation is too big for the tetrahedral hole and starts to push the anions apart. The cation-anion attractions are maintained, because cations and anions still touch, but now anion-anion repulsions are decreased because the anions no longer touch. This is an even better arrangement than the perfect fit. This situation continues to improve until *r+/r–* becomes equal to 0.414.

At this value, the cation prefers to occupy an octahedral hole because it can increase its coordination number without increasing anion-anion repulsion to any significant extent. The number of cation-anion attractive interactions increases from 4 to 6, giving a more stable lattice. As r+/r- increases between 0.414 and 0.732, cations in octahedral holes maintain their contact with 6 anions while pushing the anions further and further apart, decreasing repulsions between them. Throughout this range of *r+/r–*, occupation of an octahedral hole becomes increasingly favourable until the radius ratio becomes 0.732. The cation can now increase its coordination number to 8 anions, again increasing the number of attractive cation-anion interactions.

Radius Ratio Rules:

Range of r+/r-	Lattice Type	Example
0.224-0.414	ZnS	ZnS
0.414-0.732	NaCl CsCl	KCl CsCl
> 0.732		

Ion radii and the radius ratio rules can be used to predict the probable choice of lattice for an ionic compound. The radius ratio rules provide useful guidelines to the lattice structures of ionic compounds. It is important to realise, however, that they are not hard and fast; the predictions made using the rules are often wrong. For example, in the zincblende form of ZnS, Zn^{2+} ions occupy tetrahedral holes in a fcc lattice of S2- ions. In fact, this compound is the prototype for this type of structure. However, the radius ratio rules predict that the Zn^{2+} ions should occupy octahedral holes! The radius ratio rules fail frequently because the relative size of the ions is not the only factor that is important in determining structure. Of particular importance is the tendency towards at least some covalent bonding between cation and anion. Covalency involves directional requirements and may influence the type of hole preferred by the cation.

CRYSTAL SYSTEMS

The crystal systems are a grouping of crystal structures according to the axial system used to describe their lattice. Each crystal system consists of a set of three axes in a particular geometrical arrangement. There are seven unique crystal systems. The simplest and most symmetric, the cubic (or isometric) system, has the symmetry of a cube, that is, it exhibits four threefold rotational axes oriented at 109.5 degrees (the tetrahedral angle) with respect to each other. These threefold axes lie along the body diagonals of the cube. This definition of a cubic is correct, although many textbooks incorrectly state that a cube is defined by three mutually perpendicular axes of equal length – if this were true there would be far more than 14 Bravais lattices. The other six systems, in order of decreasing symmetry, are hexagonal, tetragonal, rhombohedral (also kn own astrigonal), or thorhombic, monoclinic and triclinic. Some crystallographers consider the hexagonal crystal system not to be its own crystal system, but instead a part of the trigonal crystal system. The crystal system and Bravais lattice of a crystal describe the (purely) translational symmetry of the crystal.

POINT AND SPACE GROUPS

The crystallographic point group or crystal class is the mathematical group comprising the symmetry operations that leave at least one point unmoved and that leave the appearance of the crystal structure unchanged. These symmetry operations can include reflection, which reflects the structure across a reflection plane, rotation, which rotates the structure a specified portion of a circle about a rotation axis, inversion which changes the sign of the coordinate of each point with respect to a centre of symmetry or inversion point and improper rotation, which consists of a rotation about an axis followed by an inversion.

Rotation axes (proper and improper), reflection planes, and centers of symmetry are collectively called symmetry elements. There are 32 possible crystal classes. Each one can be classified into one of the seven crystal systems. The space group of the crystal structure is composed of the translational symmetry operations in addition to the operations of the point group. These include pure translations which move a point along a vector, screw axes, which rotate a point around an axis while translating parallel to the axis, and glide planes, which reflect a point through a plane while translating it parallel to the plane. There are 230 distinct space groups.

LEWIS STRUCTURES

Covalent pairs of electrons Lewis called bond pairs. Atoms with valence greater than unity could possibly share more than one pair of electrons with a neighbouring atom, producing stronger multiple bonds. Electrons left over after all the valences of the atoms were satisfied, Lewis classified as lone pair electrons.

The identification of different classes of electrons in molecules proved very powerful in explaining their structures and reactivities. Molecular formulas which show the bonding and free electrons are called Lewis structure formulas. Lewis structures based on the duet/octet heuristic provide a simplified basis for explaining the properties of molecules.

ELECTRON CORRELATIONS IN SOLIDS

The variety of phenomena associated with the electronic structure of crystalline matter has been the subject of theoretical investigation for a long time. Historically, the starting impetus was given by the challenge of accounting for some peculiar properties of the metallic state, such as the excellent heat and electric conductivity. The quantum theory of electrons moving in a periodic lattice developed in the 1920's by Sommerfeld, Bloch, Wilson and others, for example, was indeed primarily aimed at understanding the electronic transport properties of metals.

This theory led to the band model of the crystalline solid. It was realized that within the band model the configuration of the valence electrons can explain the division of perfect crystals at low temperatures into metals and insulators. If the valence electrons exactly fill one or more energy bands, leaving others empty, the crystal is an insulator. The difference in energy between the top of the highest occupied band and the bottom of the lowest empty band is known as the band gap. In the other case, when partially filled bands exist, the solid has metallic properties.

The band description of solids rests, as the simple orbital theories of atoms and molecules, on the independent electron approximation. Each electron experiences the periodic field due to the atomic nuclei and an effective field obtained by averaging over the positions of the remaining electrons. In spite of this approximation band theory has been very successful in predicting and understanding the properties of real materials, including elemental metals and many (simple) covalent and ionic compounds. It has been noticed, however, already in the 1930's that for some solids the one-electron band theory is inadequate. These are systems where, due to the competition between delocalization, band-like, effects and strong (atomic-like) electron – electron interactions, the independent electron approximation breaks down. Typical examples are the 3d -metal compounds.

In these compounds strong repulsive interactions may result in electron localization and the opening of a conductivity gap even though the d-"bands"are not completely filled. The presence of bound, unpaired, 3d electrons further gives rise to local magnetic moments and very rich magnetic properties. Serious attempts to characterize and understand the electronic structure of such materials began in the late 1950's. The main motivations at the time came from the issue of the Mott metal – insulator transition, the problem of magnetic ordering, and the problem of itinerant ferromagnetism.

More recently, phenomen a like heavy fermion behaviour, high - temperature superconductivity, colossal magnet-oresistance, and spin-Peierls phase transitions have revived interest in these systems. Nevertheless, although considerable effort has been put into the field, many of the transition metal materials are poorly understood. The proper treatment of various competing physical effects, like electron localization as a result of strong electron – electron interactions and band-like behaviour as a result of orbital overlap and translational symmetry, remains one of the most difficult problems in theoretical solid state physics.

Wave-function based correlated quantum chemical methods have been successfully applied for calculating molecular electronic structures. A very attractive feature of the quantum chemical approach is the ability to address the electron correlation problem in a systematic and controlled manner. However, due to the high computational cost of these methods, inclusion of electron correlation effects in the study of crystalline solids did not seem feasible for a long time. Algorithms able to treat electron correlation explicitly in extended periodic systems have been developed only recently, either within the many-body second-order perturbation theory (MBPT2) or within the configuration interaction (CI) formalism. Still, calculations have been carried out for rather simple closed-shell systems only, like crystalline LiH and polymers.

Currently the most widely used first-principles method in solid state applications is the density functional (DF) theory within the local density approximation (LDA) (for a review). In the DF formalism, one avoids construction of the many-body wave-function and instead computes specific properties of the system from the ground-state electronic charge density. Recently, the study of the electronic response to external perturbations made possible calculation of low-energy excitations such as phonon dispersions, optical spectra and magnon dispersions.

The DF-based methods approximately include electron correlation and usually give accurate estimates for a number of physical properties — calculated crystal structures, lattice parameters, elastic constants, dielectric functions and phonon frequencies are within a few per cent of the experimental values. Notable exception is the underestimation of the band gap. For strongly correlated 3d -electron systems like CoO and La_2CuO_4, DF/LDA calculations fail actually in predicting the correct insulating ground-state.

A different approach to the problem of strongly interacting electrons in solids is the model Hamiltonian method, where many-body effects are incorporated in some parameterized form. Two of the simplest models are the Hubbard Hamiltonian widely used in a qualitative or a semi-quantitative way for studying the interplay between electron delocalization and electron repulsion effects, and the Anderson model of a magnetic impurity coupled to a conduction band.

Model Hamiltonian investigations can often describe tendencies and reveal the basic mechanisms of various physical phenomena. At the same time the model Hamiltonian approach is rather limited in its ability to make quantitative predictions, since even the simple Hubbard model can be solved exactly only in the one-dimensional case. In dealing with the two- and three- dimensional models numerical techniques are usually employed, based on quantum Monte Carlo methods or on diagonalizing the Hamiltonian for a cluster of atoms. However, the quantum Monte Carlo methods suffer from the so-called sign problem, while the exact diagonalization is limited by the exponential growth of the computational effort with the cluster size.

Local cluster (or atomic) impurity-like models have been used to describe various electronic properties of strongly correlated 3d- and 4f-metal systems for a long time, in combination with model Hamiltonians or with wave- function based quantum chemical calculations.

Within the ab initio wave-function based approach the cluster is embedded in some effective potential that accounts for the crystal Madelung field and for short-range Pauli and exchange interactions due to the finite charge distribution of the nearest neighbours, respectively. The cluster contains one or more metal ions and the adjacent ligands. Proper treatment of the many- body problem, by configuration interaction techniques or second-order perturbation theory gives in several insulating transition metal materials accurate estimates and predictions for properties like core level spectra d – d excitations ligand to metal charge transfer effects and Heisenberg spin interactions.

In the study of the local many-body physics in solids a step forward consists in solving the problem of a (single- or multi-site) quantum impurity embedded in an effective medium self-consistently. This is the essential idea behind the dynamical mean field theory (DMFT), developed during the last decade (for a review). The DMFT reduces a lattice model to an Anderson impurity subject to a self-consistently determined "hybridization" with an electron "bath" representing the crystalline host.

Its s tructure is that of aclosed set of equation sure l ating th e "correlated"single-site impurity Green function to the crystal Green function evaluated at the impurity site. The crystal Green function can be calculated using standard DF/LDA band structure methods. Within such a LDA+DMFT scheme the DMFT correction to the LDA input results in a change of the electronic charge density. This updated charge density is used in a new LDA+DMFT loop and the cycle is repeated until convergence is reached. Results of such calculations were able, for example, to describe the finite- temperature magnetic properties of iron and nickel and the phase diagram of plutonium. Efforts to extend the single-site DMFT to include multi-site, short- range correlations led to the dynamical cluster approximation (DCA) and the cellular dynamical field theory (CDMFT).

BAND STRUCTURE OF ELECTRONIC

The electronic band structure (or simply band structure) of a solid describes ranges of energy that an electron is "forbidden" or "allowed" to have. It is due to the diffraction of the quantum mechanical electron waves in the periodic crystal lattice. The band structure of a material determines several characteristics, in particular its electronic and optical properties.

BANDS OCCUR IN MATERIALS

The electrons of a single free-standing atom occupy atomic orbitals, which form a discrete set of energy levels. If several atoms are brought together into a molecule, their atomic orbitals split, as in a coupled oscillation. This produces a number of molecular orbitals proportional to the number of atoms.

When a large number of atoms (of order 1020 or more) are brought together to form a solid, the number of orbitals becomes exceedingly large, and the difference in energy between them becomes very small, so the levels may be considered to form continuous bands of energy rather than the discrete energy levels of the atoms in isolation. However, some intervals of energy contain no orbitals, no matter how many atoms are aggregated, forming band gaps.

Within an energy band, energy levels are so numerous as to be a near continuum. First, the separation between energy levels in a solid is comparable with the energy that electrons constantly exchange with phonons (atomic vibrations). Second, it is comparable with the energy uncertainty due to the Heisenberg uncertainty principle, for reasonably long intervals of time. As a result, the separation between energy levels is of no consequence.

Basic Concepts

Any solid has a large number of bands. In theory, it can be said to have infinitely many bands. However, all but a few lie at energies so high that any electron that reaches those energies escapes from the solid. These bands are usually disregarded.Bands have different widths, based upon the properties of the atomic orbitals from which they arise. Also, allowed bands may overlap, producing (for practical purposes) a single large band.Metals contain a band that is partly empty and partly filled regardless of temperature.

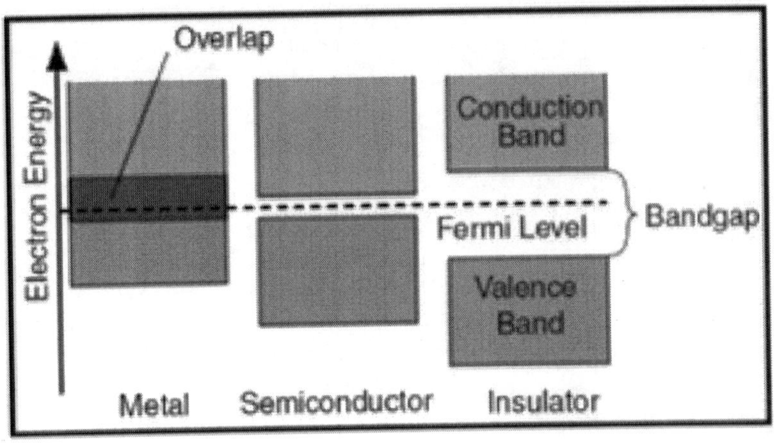

Fig. Electronic Band Structure of Metals

Therefore they have very high conductivity.The lowermost, almost fully occupied band in an insulator or semiconductor is called the valence band by analogy with the valence electrons of individual atoms. The uppermost, almost unoccupied band is called the conduction band because only when electrons are excited to the conduction band can current flow in these materials.

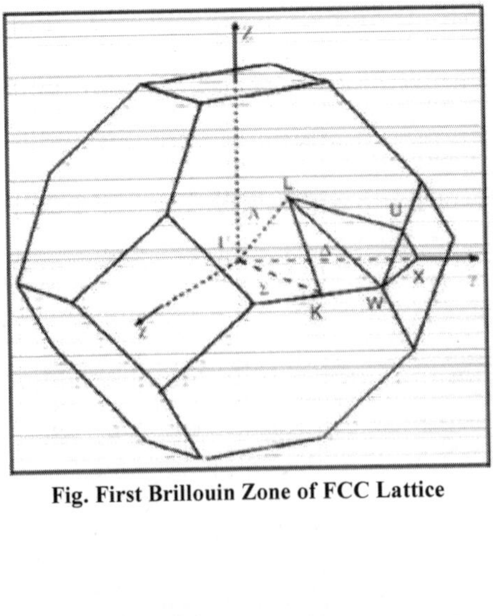

Fig. First Brillouin Zone of FCC Lattice

The difference between insulators and semiconductors is only that theforbidden band gap between the valence band and conduction band is larger in an insulator, so that fewer electrons are found there and the electrical conductivity is lower.

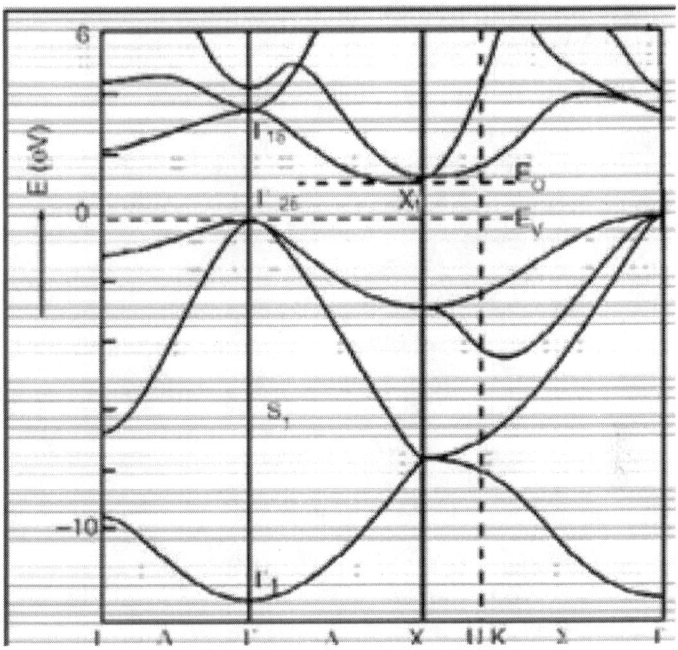

Fig. Band Structure in Silicon

Because one of the main mechanisms for electrons to be excited to the conduction band is due to thermal energy, the conductivity of semiconductors is strongly dependent on the temperature of the material. This band gap is one of the most useful aspects of the band structure, as it strongly influences the electrical and optical properties of the material. Electrons can transfer from one band to the other by means of carrier generation and recombination processes. The band gap and defect states created in the band gap by doping can be used to create semiconductor devices such as solar cells, diodes, transistors, laser diodes, and others.

Symmetry

A more complete view of the band structure takes into account the periodic nature of a crystal lattice using the symmetry operations that form a space group. The Schrödinger equation is solved for the crystal, which has Bloch waves as solutions:

$$\psi_{nk}(r) = e^{ik.r} u_{nk}(r),$$

where k is called the wavevector, and is related to the direction of motion of the electron in the crystal, and n is the band index, which simply numbers theenergy bands. The wave vectork takes on values within the Brillouin zone (BZ) corresponding to the crystal lattice, and particular directions/points in the BZ are assigned conventional names like $T\Delta \wedge \sum$ etc. These directions are shown for the face-centered cubic lattice geometry in.

The available energies for the electron also depend upon k, as shown in Figure for silicon in the more complex energy band diagram at the right. In this diagram the topmost energy of the valence band is labeled Ev and the bottom energy in the conduction band is labeled Ec.

The top of the valence band is not directly below the bottom of the conduction band (Ev is for an electron traveling in direction G, Ec in direction X), so silicon is called an indirect gap material.

For an electron to be excited from the valence band to the conduction band, it needs something to give it energy Ec "Ev and a change in direction/ momentum. In other semiconductors (for example GaAs) both are at G, and these materials are called direct gap materials (no momentum change required). Direct gap materials benefit the operation of semiconductor laser diodes. Anderson's rule is used to align band diagrams between two different semiconductors in contact.

Band Structures in Different Types of Solids

Although electronic band structures are usually associated with crystalline materials, quasi-crystalline and amorphous solids may also exhibit band structures. However, the periodic nature and symmetrical properties of crystalline materials makes it much easier to examine the band structures of these materials theoretically. In addition, the well-defined symmetry axes of crystalline materials makes it possible to determine the dispersion relationship between the momentum (a 3-dimension vector quantity) and energy of a material. As a result, virtually all of the existing theoretical work on the electronic band structure of solids has focused on crystalline materials.

Density of States

While the density of energy states in a band could be very large for some materials, it may not be uniform. It approaches zero at the band boundaries, and is generally highest near the middle of a band. The density of states for the free electron model in three dimensions is given by,

$$D(\hat{I}) = \frac{V}{2\pi^2}\left(\frac{2m}{h^2}\right)^{3/2} \in^{1/2}$$

Filling of Bands

Although the number of states in all of the bands is effectively infinite, in an uncharged material the number of electrons is equal only to the number of protons in the atoms of the material.

Therefore not all of the states are occupied by electrons ("filled") at any time. The likelihood of any particular state being filled at any temperature is given by the Fermi-Dirac statistics. The probability is given by the following:

$$f(E)=\frac{1}{1+e\frac{E-E_F}{k_B T}}$$

where:
k_B is Boltzmann's constant,
T is the temperature,
E_F is the Fermi energy (or 'Fermi level').

The Fermi level naturally is the level at which the electrons and protons are balanced.

At $T = 0$, the distribution is a simple step function:

$$f(E)=\begin{cases}1 & if\, 0<E\leq E_F\\ 0 & if E_F<E\end{cases}$$

At non-zero temperatures, the step "smooths out", so that an appreciable number of states below the Fermi level are empty, and some states above the Fermi level are filled.

BAND STRUCTURE OF CRYSTALS

BRILLOUIN ZONE

Because electron momentum is the reciprocal of space, the dispersion relation between the energy and momentum of electrons can best be described in reciprocal space. It turns out that for crystalline structures, the dispersion relation of the electrons is periodic, and that the Brillouin zone is the smallest repeating space within this periodic structure. For an infinitely large crystal, if the dispersion relation for an electron is defined throughout the Brillouin zone, then it is defined throughout the entire reciprocal space.

THEORY OF BAND STRUCTURES

The ansatz is the special case of electron waves in a periodic crystal lattice using Bloch waves as treated generally in the dynamical theory of diffraction. Every crystal is a periodic structure which can be characterized by a Bravais lattice, and for each Bravais lattice we can determine the reciprocal lattice, which encapsulates the periodicity in a set of three reciprocal lattice vectors (b_1, b_2, b_3). Now, any periodic potential $V(r)$ which shares the same periodicity as the direct lattice can be expanded out as a Fourier series whose only non-vanishing components are those associated with the reciprocal lattice vectors.

So the expansion can be written as:

$$V(r) = \sum_K V_{k^{eoL.r}}$$

where $K = m_1 b_1 + m_2 b_2 + m_3 b_3$ for any set of integers (m_1, m_2, m_3).

From this theory, an attempt can be made to predict the band structure of a particular material, however most ab initio methods for electronic structure calculations fail to predict the observed band gap.

Nearly-free Electron Approximation

In the nearly-free electron approximation in solid state physics interactions between electrons are completely ignored. This approximation allows use of Bloch's Theorem which states that electrons in a periodic potential have wavefunctions and energies which are periodic in wavevector up to a constant phase shift between neighbouring reciprocal lattice vectors.

The consequences of periodicity are described mathematically by the Bloch wavefunction:

$$Y_{m,k}(r) = e^{ik.r} u_n(r)$$

where the function $u_n(r)$ is periodic over the crystal lattice, that is,

$$u_n(r) = u_n(r - R).$$

Here index n refers to the n-th energy band, wavevector k is related to the direction of motion of the electron, r is position in the crystal, and R is location of an atomic site.

Tight-binding Model

The opposite extreme to the nearly-free electron approximation assumes the electrons in the crystal behave much like an assembly of constituent atoms. This tight-binding model assumes the solution to the time-independent single electron Schrödinger equation Ø is well approximated by a linear combination of atomic orbitals ψ *n*(*r*).

$$Y(r) = \sum_{n,R} b_{n,R} \psi_n (r - R),$$

where the coefficients *bn,R* are selected to give the best approximate solution of this form. Index n refers to an atomic energy level and R refers to an atomic site. A more accurate approach using this idea employs Wannier functions, defined by,

$$a_n(r - R) = \frac{V_c}{(2\pi)^3} \int_{BZ} dk e^{-ik.(R-r)_{unk;}}$$

in which *unk* is the periodic part of the Bloch wave and the integral is over the Brillouin zone. Here index n refers to the n-th energy band in the crystal. The Wannier functions are localized near atomic sites, like atomic orbitals, but being defined in terms of Bloch functions they are accurately related to solutions based upon the crystal potential. Wannier functions on different atomic sites R are orthogonal. The Wannier functions can be used to form the Schrödinger solution for the n-th energy band as:

$$Y_{n,k}(r) = \sum_{R} e^{-ik.(R-r)_{a_n(r-R)}}$$

Ab initio Density-functional Theory

In present days physics literature, the large majority of the electronic structures and band plots is calculated using the density-functional theory (DFT) which is not a model but rather an ab initio theory, *i.e.* a microscopic first-principle theory of condensed matter physics that tries to cope with the electron-electron many-body problem via the introduction of an exchange- correlation term in the functional of the electronic density. DFT calculated bands are found in many cases in agreement with experimental measured bands, for example by angle-resolved photoemission spectroscopy (ARPES). In particular, the band shape seems well reproduced by DFT. But also there are systematic errors of DFT bands with respect to the experiment. In particular, DFT seems to underestimate systematically by a 30-40 per cent the band gap in insulators and semiconductors. It must be said that DFT is in principle an exact theory to reproduce and predict ground state properties (*e.g.* the total energy, the atomic structure, etc.).

However DFT is not a theory to address excited state properties, such as the band plot of a solid that represents the excitation energies of electrons injected or removed from the system. What in literature is quoted as a DFT band plot is a representation of the DFT Kohn-Sham energies, that is the energies of a fictive non-interacting system, the Kohn-Sham system, which has no physical interpretation at all.

The Kohn-Sham electronic structure must not be confused with the real, quasiparticle electronic structure of a system, and there is no Koopman's theorem holding for Kohn-Sham energies, like on the other hand for Hartree- Foc k e nergies tha t ca n be truly con side red a s an a pprox im ation for quasiparticle energies. Hence in principle DFT is not a band theory, not a theory suitable to calculate bands and band-plots.

GREEN'S FUNCTION

To calculate the bands including electron-electron interaction many-body effects, one can resort to so called Green's function methods. Indeed, the knowledge of the Green's function of a system provides both ground (the total energy) and also excited state observables of the system. The poles of the Green's function are the quasiparticle energies, the bands of a solid. The Green's function can be calculated by solving the Dyson equation once the self-energy of the system is known. For real systems like solids, the self-energy is a very complex quantity and usually approximations are needed to solve the problem.

One of such approximations is the GW approximation, so called from the mathematical form the self-energy takes as product Ó = GW of the Green's function G and the dynamically screened interaction W. This approach is more pertinent to address the calculation of band plots (and also quantities beyond, such as the spectral function) and can be also formulated in a completely ab initio way. The GW approximation seems to provide band gaps of insulators and semiconductors in agreement with the experiment and hence to correct the systematic DFT underestimation.

Mott Insulators

Although the nearly-free electron approximation is able to describe many properties of electron band structures, one consequence of this theory is that it predicts the same number of electrons in each unit cell. If the number of electrons is odd, we would then expect that there is an unpaired electron in each unit cell, and thus that the valence band is not fully occupied, making the material a conductor.

However, materials such as CoO that have an odd number of electrons per unit cell are insulators, in direct conflict with this result. This kind of material is known as a Mott insulator, and requires inclusion of detailed electron-electron interactions (treated only as an averaged effect on the crystal potential in band theory) to explain the discrepancy. The Hubbard model is an approximate theory that can include these interactions.

CRYSTALLOGRAPHY AND CRYSTAL DEFECTS

A plane of atoms can glide rigidly over its neighbour in process described as slip with the slip system defined by the plane and the direction of slip. This kind of deformation requires enormous stresses, far greater than those required to actually deform a crystal. This is because almost all crystals contain defects known as dislocations. A dislocation enables the planes to glide in a piecewise manner rather than the rigid displacement of the entire plane. This greatly reduces the stress required to cause slip. A good analogy to illustrate the role of a dislocation is to imagine the force required to pull an entire carpet along the floor. On the other hand, if a bump is introduced into the carpet, the force needed to move the bump along is much smaller.

The amount and direction of displacement produced by propagating the bump is, in the context of a dislocation, known as its Burgers vector.

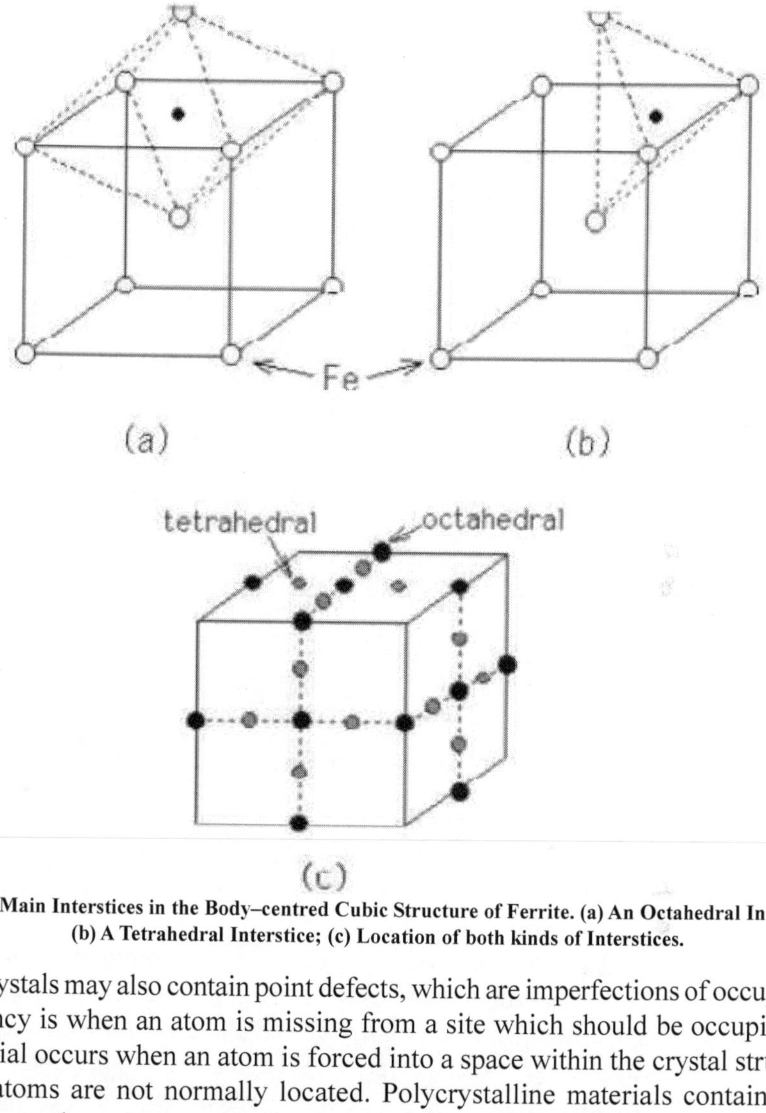

Fig. The Main Interstices in the Body–centred Cubic Structure of Ferrite. (a) An Octahedral Interstice; (b) A Tetrahedral Interstice; (c) Location of both kinds of Interstices.

Crystals may also contain point defects, which are imperfections of occupation. A vacancy is when an atom is missing from a site which should be occupied. An interstitial occurs when an atom is forced into a space within the crystal structure, where atoms are not normally located. Polycrystalline materials contain many crystals; another common term for crystals in such materials is grains. Atoms in the grain boundary between crystals must in general be displaced from positions they would occupy in the undisturbed crystal. Therefore, grain boundaries are defects.

CRYSTALS AND X-RAYS

Crystals have long been admired for their regularity and symmetry, but they were not investigated scientifically until the 17th century. Johannes Kepler hypothesized in his work Strena seu de Nive Sexangula that the hexagonal symmetry of snowflake crystals was due to a regular packing of spherical water particles. As shown by X-ray crystallography, the hexagonal symmetry of snowflakes results from the tetrahedral arrangement of hydrogen bonds about each water molecule.

The water molecules are arranged simiarly to the silicon atoms in the tridymite polymorph of SiO2. The resulting crystal structure has hexagonal symmetry when viewed along a principal axis.

Crystal symmetry was first investigated experimentally by Nicolas Steno who showed that the angles between the faces are the same in every exemplar of a particular type of crystal who discovered that every face of a crystal can be described by simple stacking patterns of blocks of the same shape and size. Hence, William Hallowes Miller in 1839 was able to give each face a unique label of three small integers, the Miller indices which are still used today for identifying crystal faces. Haüy's study led to the correct idea that crystals are a regular three-dimensional array (a Bravais lattice) of atoms and molecules; a single unit cell is repeated indefinitely along three principal directions that are not necessarily perpendicular.

In the 19th century, a complete catalog of the possible symmetries of a crystal was worked out by Johann Hessel, Auguste Bravais, Yevgraf Fyodorov,, Arthur Schönflies and (belatedly) William Barlow. On the basis of the available data and physical reasoning, Barlow proposed several crystal structures in the 1880s that were validated later by X-ray crystallography; however, the available data were too few in the 1880s to accept his models as conclusive.

X-rays were discovered by Wilhelm Conrad Röntgen in 1895, just as the studies of crystal symmetry were being concluded. Physicists were initially uncertain of the nature of X-rays, although it was soon suspected (correctly) that they were waves of electromagnetic radiation, in other words, another form of light. At that time, the wave model of light — specifically, the Maxwell theory of electromagnetic radiation — was well accepted among scientists, and experiments by Charles Glover Barkla showed that X-rays exhibited phenomena associated with electromagnetic waves, including transverse polarization an d spe ctral lines a kin to thos e obs erved in th e v isible wavelengths.

Single-slit experiments in the laboratory of Arnold Sommerfeld suggested the wavelength of X-rays was roughly 1 Angström, one ten millionth of a millimetre. However, X-rays are composed of photons, and thus are not only waves of electromagnetic radiation but also exhibit particle-like properties. The photon concept was introduced by Albert Einstein in 1905, but it was not broadly accepted until 1922, when Arthur Compton confirmed it by the scattering of X-rays from electrons. Therefore, these particle-like properties of X-rays, such as their ionization of gases, caused William Henry Bragg to argue in 1907 that X-rays were not electromagnetic radiation. Nevertheless, Bragg's view was not broadly accepted and the observation of X-ray diffraction in 1912 confirmed for most scientists that X-rays were a form of electromagnetic radiation.

X-RAYS ANALYSIS

The incoming beam (coming from upper left) causes each scatterer to re- radiate a small portion of its energy as a spherical wave. If scatterers are arranged symmetrically with a separation d, these spherical waves will be in synch (add constructively) only in directions where their path-length difference 2d sin q equals an integer multiple of the wavelength l. In that case, part of the incoming beam is deflected by an angle 2l, producing a reflection spot in the diffraction pattern.

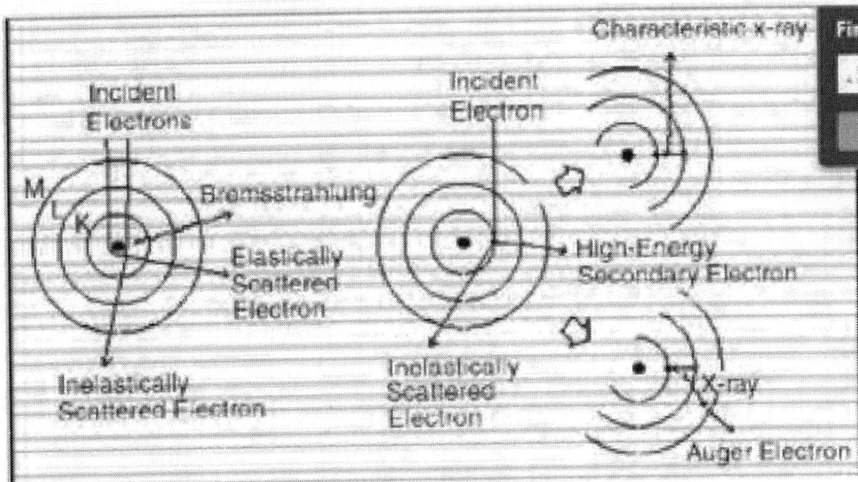

Fig. Spectrum Analysis of X-ray

Crystals are regular arrays of atoms, and X-rays can be considered waves of electromagnetic radiation. Atoms scatter X-ray waves, primarily through the atoms' electrons. Just as an ocean wave striking a lighthouse produces secondary circular waves emanating from the lighthouse, so an X-ray striking an electron produces secondary spherical waves emanating from the electron. This phenomenon is known as scattering, and the electron (or lighthouse) is known as the scatterer. A regular array of scatterers produces a regular array of spherical waves. Although these waves cancel one another out in most directions (destructive interference), they add constructively in a few specific directions, determined by Bragg's law

$$2d \sin\theta = n\lambda$$

where n is any integer. These specific directions appear as spots on the diffraction pattern, often called reflections. Thus, X-ray diffraction results from an electromagnetic wave (the X-ray) impinging on a regular array of scatterers (the repeating arrangement of atoms within the crystal).

X -ra ys are used to produc e the diffrac tion pa ttern be ca use th eir wavelength λ is typically the same order of magnitude (1-100 Ångströms) as the spacing d between planes in the crystal. In principle, any wave impinging on a regular array of scatterers produces diffraction, as predicted first by Francesco Maria Grimaldi in 1665.

To produce significant diffraction, the spacing between the scatterers and the wavelength of the impinging wave should be roughly similar in size. For illustration, the diffraction of sunlight through a bird's feather was first reported by James Gregory in the later 17th century. The first man-made diffraction gratings for visible light were constructed by David Rittenhouse in 1787, and Joseph von Fraunhofer in 1821. However, visible light has too long a wavelength (typically, 5500 Ångströms) to observe diffraction from crystals.

However, prior to the first X-ray diffraction experiments, the spacings between lattice planes in a crystal were not known with certainty. The idea that crystals could be used as a diffraction grating for X-rays arose in 1912 in a conversation between Paul Peter Ewald and Max von Laue in the English Garden in Munich. Ewald had proposed a resonator model of crystals for his thesis, but this model could not be validated using visible light, since the wavelength was much larger than the spacing between the resonators.

Von Laue realized that electromagnetic radiation of a shorter wavelength was needed to observe such small spacings, and suggested that X-rays might have a wavelength comparable to the unit-cell spacing in crystals. Von Laue worked with two technicians, Walter Friedrich and his assistant Paul Knipping, to shine a beam of X-rays through a sphalerite crystal and record its diffraction on a photographic plate.

After being developed, the plate showed a large number of well-defined spots arranged in a pattern of intersecting circles around the spot produced by the central beam. Von Laue developed a law that connects the scattering angles and the size and orientation of the unit-cell spacings in the crystal, for which he was awarded the Nobel Prize in Physics in 1914.

As described in the mathematical derivation below, the X-ray scattering is determined by the density of electrons within the crystal. Since the energy of an X-ray is much greater than that of an atomic electron, the scattering may be modeled as Thomson scattering, the interaction of an electromagnetic ray with a free electron.

This model is generally adopted to describe the polarization of the scattered radiation. The intensity of Thomson scattering declines as $1/m^2$ with the mass m of the charged particle that is scattering the radiation; hence, the atomic nuclei, which are thousands of times heavier than an electron, contribute negligibly to the scattered X-rays.

X-RAY CRYSTALLOGRAPHY

In 1912, von Laue predicted that crystals could serve as diffraction gratings if the wavelength of radiation used were approximately the same as the distance between planes of atoms in the crystals. This spacing was known to be on the order of 1 × 10-10 m, corresponding to wavelengths in the x-ray region of the electromagnetic spectrum. Subsequently, WH Bragg discovered that when X-rays of appropriate wavelength are directed onto the smooth face of a crystal, the X-rays are diffracted and a pattern characteristic of the lattice structure of the crystal is obtained. This observation is the foundation for the modern technique of X-ray

Crystallography, the most powerful method that we have for determining molecular structure. X-ray crystallography is based on the interaction of light with matter. Unlike spectroscopy, however, the light is not absorbed by the matter in this case. Instead, it is scattered from the atoms of the crystal in a very definite way. Analysis of the scattering enables the arrangements of atoms within the crystal to be deduced.

The analysis of x-ray diffraction patterns is complex, even for materials of simple structure. However, thanks to the efforts of many crystallographers over many years, it is today done quite quickly and routinely. Once a suitable crystal of a substance is grown, the modern crystallographer is able to mount it in a fully automated x-ray diffractometer, the instrument that is used for the x-ray diffraction study. The diffractometer automatically obtains all of the necessary data for the determination of the lattice structure.

The data are then fed to one of a number of computer programmes that have been developed and refined for the analysis of such data, and the structure is "solved." The output of the programme reveals not only the lattice structure, but also the locations of all of the atoms of a molecule of the substance. Structures of substances ranging from simple salts or molecular substances to huge complex proteins and enzymes can be determined using this approach. The structures of two molecules, obtained from x-ray diffraction data.

PRODUCTION OF X-RAYS

X-rays are electromagnetic radiation with wavelengths between about 0.02 Å and 100 Å (1Å = 10-10 meters).

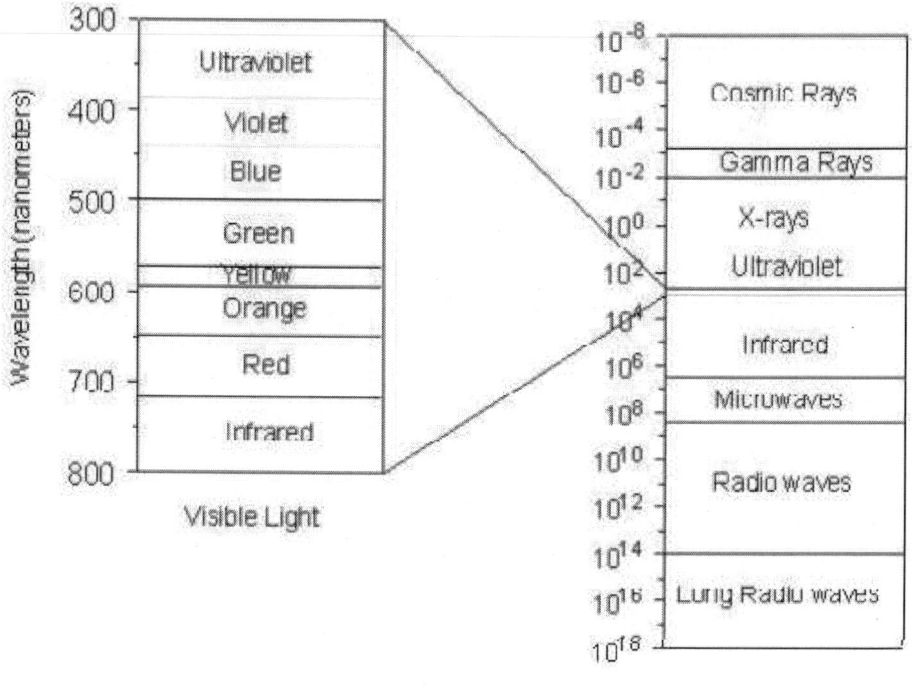

They are part of the electromagnetic spectrum that includes wavelengths of electromagnetic radiation called visible light which our eyes are sensitive to (different wavelengths of visible light appear to us as different colours). Because X-rays have wavelengths similar to the size of atoms, they are useful to explore within crystals.

The energy of X-rays, like all electromagnetic radiation, is inversely proportional to their wavelength as given by the Einstein equation:

E = h = hc

where: E = energy

h = Planck's constant, 6.62517 × 10-27 erg.sec

= frequency

c = velocity of light = 2.99793 × 1010 cm/sec

= wavelength

Thus, since X-rays have a smaller wavelength than visible light, they have higher energy. With their higher energy, X-rays can penetrate matter more easily than can visible light. Their ability to penetrate matter depends on the density of the matter, and thus X-rays provide a powerful tool in medicine for mapping internal structures of the human body.

X-rays are produced in a device called an X-ray tube. Such a tube is illustrated here. It consists of an evacuated chamber with a tungsten filament at one end of the tube, called the cathode, and a metal target at the other end, called an anode. Electrical current is run through the tungsten filament, causing it to glow and emit electrons. A large voltage difference (measured in kilovolts) is placed between the cathode and the anode, causing the electrons to move at high velocity from the filament to the anode target.

Upon striking the atoms in the target, the electrons dislodge inner shell electrons resulting in outer shell electrons having to jump to a lower energy shell to replace the dislodged electrons. These electronic transitions results in the generation of X-rays. The X-rays then move through a window in the X- ray tube and can be used to provide information on the internal arrangement of atoms in crystals or the structure of internal body parts.

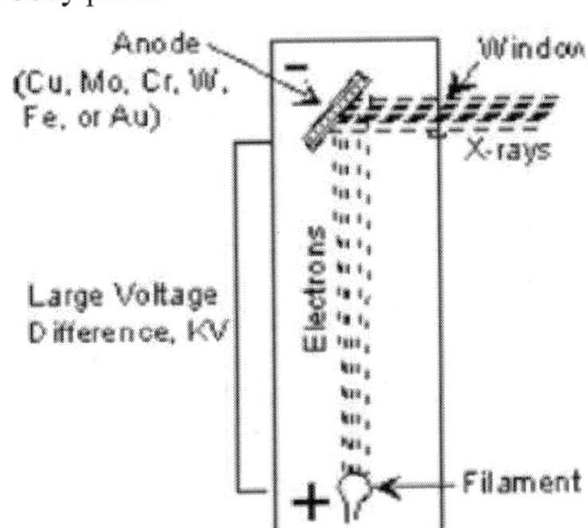

CHARACTERISTIC X-RAY SPECTRA

When the target material of the X-ray tube is bombarded with electrons accelerated from the cathode filament, two types of X-ray spectra are produced. The first is called the continuous spectra. The continuous spectra consists of a range of wavelengths of X-rays with minimum wavelength and intensity (measured in counts per second) dependent on the target material and the voltage across the X-ray tube.

The minimum wavelength decreases and the intensity increases as voltage increases. The second type of spectra, called the characteristic spectra, is produced at high voltage as a result of specific electronic transitions that take place within individual atoms of the target material.

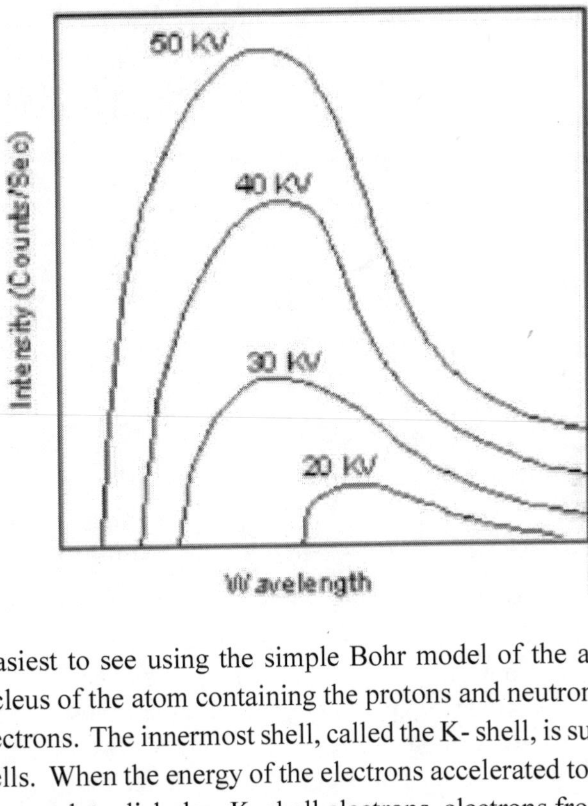

This is easiest to see using the simple Bohr model of the atom. In such a model, the nucleus of the atom containing the protons and neutrons is surrounded by shells of electrons. The innermost shell, called the K- shell, is surrounded by the L- and M - shells. When the energy of the electrons accelerated towards the target becomes high enough to dislodge K- shell electrons, electrons from the L - and M - shells move in to take the place of those dislodged.

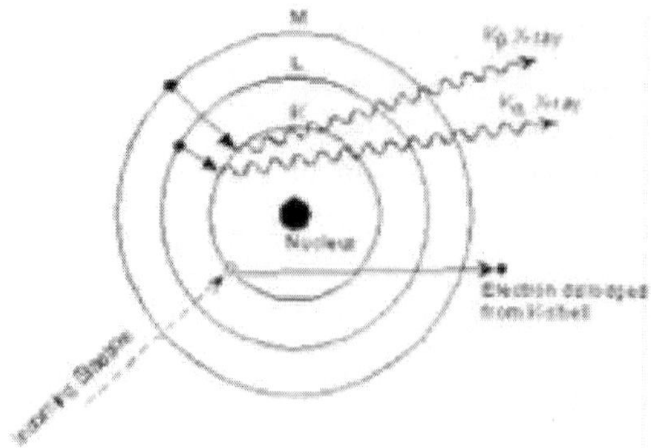

Each of these electronic transitions produces an X-ray with a wavelength that depends on the exact structure of the atom being bombarded. A transition from the L - shell to the K- shell produces a K X-ray, while the transition from an M - shell to the K- shell produces a K X-ray. These characteristic X-rays have a much higher intensity than those produced by the continuous sprectra, with K X-rays having higher intensity than K X-rays.

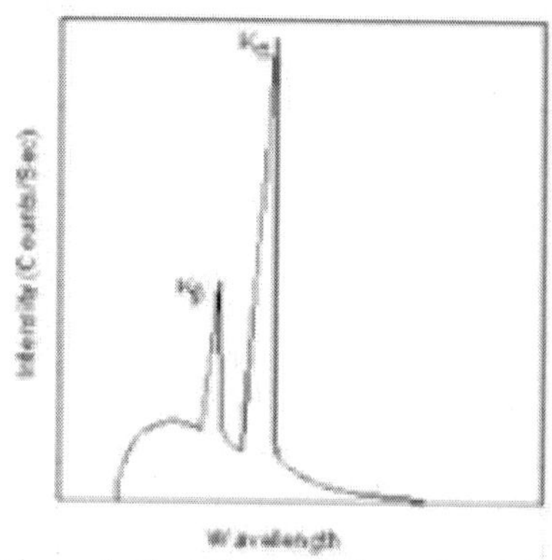

The important point here is that the wavelength of these characteristic x- rays is different for each atom in the periodic table (of course only those elements with higher atomic number have L– and M– shell electrons that can undergo transitions to produce X-rays). A filter is generally used to filter out the lower intensity K X-rays. For commonly used target materials in X-ray tubes, the X-rays have the following well-known experimentally determined wavelengths:

Element	K Wavelength () Å
Mo	0.7107
Cu	1.5418
Co	1.7902
Fe	1.9373
Cr	2.2909

BRAGG'S LAW

Since a beam of X-rays consists of a bundle of separate waves, the waves can interact with one another. Such interaction is termed interference. If all the waves in the bundle are in phase, that is their crests and troughs occur at exactly the same position (the same as being an integer number of wavelengths out of phase, $n = 1, 2, 3, 4$, etc.), the waves will interfere with one another and their amplitudes will add together to produce a resultant wave that is has a higher amplitude (the sum of all the waves that are in phase.

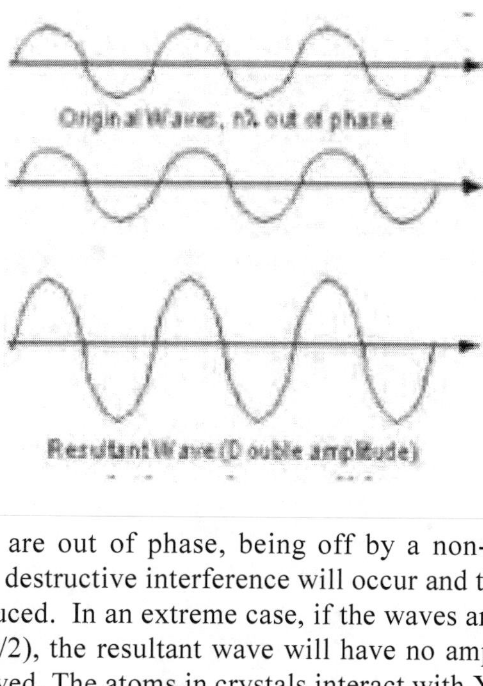

If the waves are out of phase, being off by a non-integer number of wavelengths, then destructive interference will occur and the amplitude of the waves will be reduced. In an extreme case, if the waves are out of phase by a multiple of 1/2 ($n/2$), the resultant wave will have no amplitude and thus be completely destroyed. The atoms in crystals interact with X-ray waves in such a way as to produce interference. The interaction can be thought of as if the atoms in a crystal structure reflect the waves. But, because a crystal structure consists of an orderly arrangement of atoms, the reflections occur from what appears to be planes of atoms.

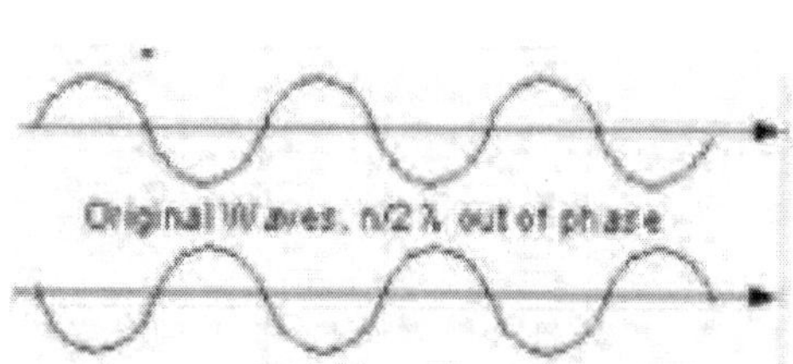

Let's imagine a beam of X-rays entering a crystal with one of these planes of atoms oriented at an angle of to the incoming beam of monochromatic X-rays (monochromatic means one colour, or in this case 1 discreet wavelength as produced by the characteristic spectra of the X-ray tube). Two such X-rays are shown here, where the spacing between the atomic planes occurs over the distance, d. Ray 1 reflects off of the upper atomic plane at an angle ? equal to its angle of incidence. Similarly, Ray 2 reflects off the lower atomic plane at the same angle.

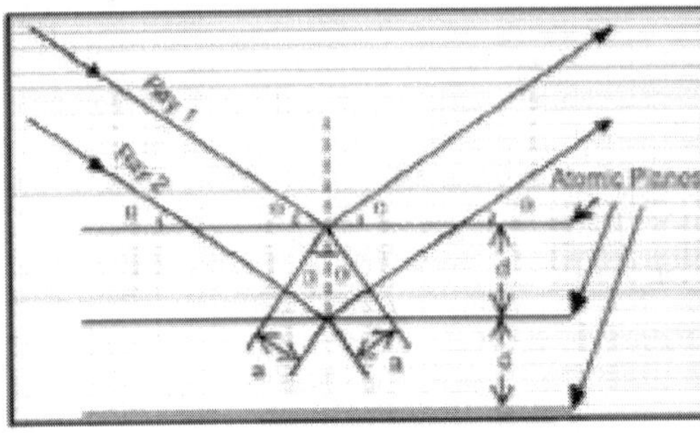

While Ray 2 is in the crystal, however, it travels a distance of 2a farther than Ray 1. If this distance 2a is equal to an integral number of wavelengths (n?), then Rays 1 and 2 will be in phase on their exit from the crystal and constructive interference will occur. If the distance 2a is not an integral number of wavelengths, then destructive interference will occur and the waves will not be as strong as when they entered the crystal.

Thus, the condition for constructive interference to occur is, $n = 2a$ but, from trigonometry, we can figure out what the distance 2a is in terms of the spacing, d, between the atomic planes.

$$a = d \sin \text{ or } 2a = 2d \sin$$

thus,

$$n = 2\delta \sin$$

This is known as Bragg's Law for X-ray diffraction. What it says is that if we know the wavelength ,l, of the X-rays going in to the crystal, and we can measure the angle q of the diffracted X-rays coming out of the crystal, then we know the spacing (referred to as d-spacing) between the atomic planes.

$$d = \frac{n}{2\sin}$$

Again it is important to point out that this diffraction will only occur if the rays are in phase when they emerge, and this will only occur at the appropriate value of n (1, 2, 3, etc.) and θ.

In theory, then we could re-orient the crystal so that another atomic plane is exposed and measure the d-spacing between all atomic planes in the crystal, eventually leading us to determine the crystal structure and the size of the unit cell.

X-RAY METHOD

In practice, this would be a time consuming operation to reorient the crystal, measure the angle q, and determine the d-spacing for all atomic planes. A faster way is to use a method called the powder method. In this method, a mineral is ground up to a fine powder. In the powder, are thousands of grains that have random orientations. With random orientations we might expect most of the different atomic planes to lie parallel to the surface in some of the grains.

Thus, by scanning through an angle q of incident X-ray beams form 0 to 90°, we would expect to find all angles where diffraction has occurred, and with a different atomic spacing.

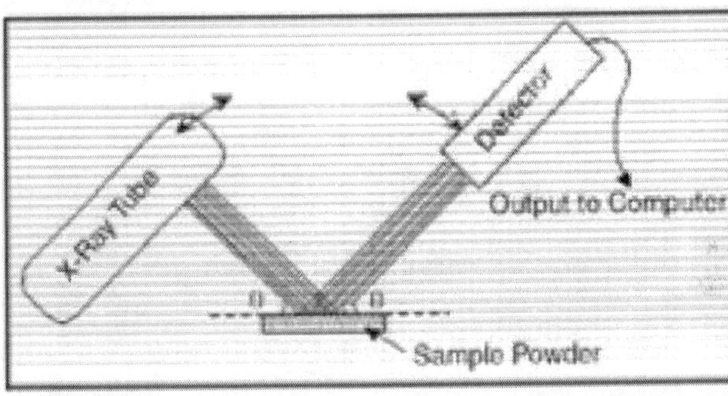

The instrument used to do this is an x-ray powder diffractometer. It consists of an X-ray tube capable of producing a beam of monochromatic X- rays that can be rotated to produce angles from 0 to 90°. A powdered mineral sample is placed on a sample stage so that it can be irradiated by the X-ray tube. To detect the diffracted X-rays, an electronic detector is placed on the other side of the sample from the X-ray tube, and it too is allowed to rotate to produce angles from 0 to 90°.

The instrument used to rotate both the X-ray tube and the detector is called a goniometer. The goniometer keeps track of the angle and sends this information to a computer, while the detector records the rate of X-rays coming out the other side of the sample (in units of counts/sec) and sends this information to the computer.

After a scan of the sample the X-ray intensity can be plotted against the angle (usually reported as 2 because of the way older diffractometers were made) to produce a chart, like the one shown here. The angle 2 for each diffraction peak can then be converted to d-spacing, using the Bragg equation.

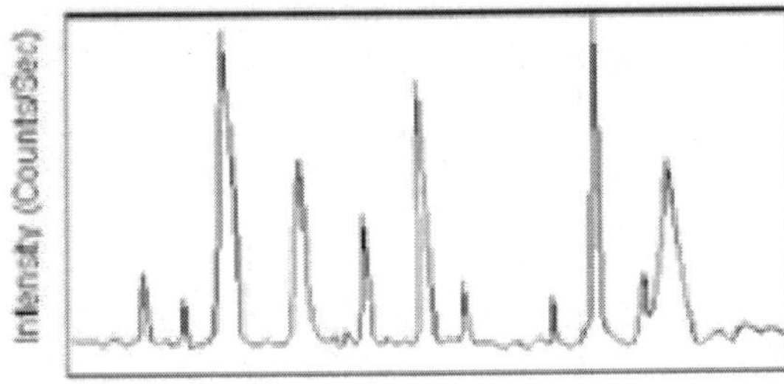

One can then work out the crystal structure and associate each of the diffraction peaks with a different atomic plane in terms of the Miller Index for that plane (hkl). A group known as the Joint Committee on Powder Diffraction Standards (JCPDS) has collected data such as this on thousands of crystalline substances. This data can be obtained as the JCPDS Powder Diffraction File. Since every compound with the same crystal structure will produce an identical powder diffraction pattern, the pattern serves as kind of a "fingerprint" for the substance, and thus comparing an unknown mineral to those in the Powder Diffraction file enables easy identification of the unknown.

❐

2.

Quantum Theory

Quantum chemistry is based on the postulates of Quantum Mechanics. The theory of quantum chemistry with an emphasis on their practical implications for the molecular modeller, and we will try to answer the questions of when to use a quantum chemical method instead of molecular mechanics, which quantum chemical method to choose, and what to expect from the quality of the results. In quantum chemistry, the system is described by a wavefunction which can be found by solving the Schrödinger equation. This equation relates the stationary states of the system and their energies to the Hamiltonian operator, which can be viewed as the recipe for obtaining the energy associated with a wavefunction describing the positions of the nuclei and electrons in the system. In practice the Schrödinger equation cannot be solved exactly and approximations have to be made.

The approach is called "ab initio" when it makes no use of empirical information, except for the fundamental constants of nature such as the mass of the electron, Planck's constant etc., that are required to arrive at numerical predictions. Do not confuse "ab initio" with "exact" ! In spite of the necessary approximations, ab initio theory has the conceptual advantage of generality, and the practical advantage that (with experience) its successes and failures are more or less predictable.

Aspects of molecular structure and dynamics can be modelled using classical methods in the form of molecular mechanics and dynamics. The classical force field is based on empirical results, averaged over a large number of molecules. Because of this extensive averaging, the results can be good for standard systems, but there are many important questions in chemistry that can not at all be addressed by means of this empirical approach. If one wants to know more than just structure or other properties that are derived only from the potential energy surface, in particular properties that depend directly on the electron density distribution, one has to resort to a more fundamental and general approach: quantum chemistry. The same holds for all non- standard cases for which molecular mechanics is simply not applicable. The major disadvantage of ab initio quantum chemistry are the heavy demands on computer power.

Therefore, further approximations have been applied for a long time which go together with the introduction of empirical parameters into the theoretical model. This has led to a number of semi-empirical quantum chemical methods, which can be applied to larger systems, and give reasonable electronic wavefunctions so that electronic properties can be predicted. Compared with ab initio calculations their reliability is less and their applicability is limited by the requirement for parameters, just like in molecular mechanics.

In general, one should apply quantum chemistry for "small" systems, which can be treated at a very high level, when electronic properties are sought (electric moments, polarizabilities, shielding constants in NMR and ESR, etc.) and for "non-standard" structures, for which no valid molecular mechanics parameters are available. Examples are conjugated pi systems, organometallic compounds and other systems with unusual bond or atom types, excited states, reactive intermediates, and generally structures with unusual electronic effects.

THEORY OF QUANTUM CHROMODYNAMICS

In theoretical physics, quantum chromodynamics (QCD) is a theory of the strong interaction (colour force), a fundamental force describing the interactions between quarks and gluons which make up hadrons (such as the proton, neutron or pion). It is the study of the SU(3) Yang–Mills theory of colour-charged fermions (the quarks). QCD is a quantum field theory of a special kind called a non-abelian gauge theory, consisting of a 'colour field' mediated by a set of exchange particles (the gluons). The theory is an important part of the Standard Model of particle physics. A huge body of experimental evidence for QCD has been gathered over the years.

The standard model postulates that all known particles are either fundamental point particles or are composed of fundamental point particles according to a remarkably small set of rules. Just as atoms are bound states of atomic nuclei and electrons, atomic nuclei are bound states of protons and neutrons.

We delve one step deeper in the hierarchy of the universe. We now believe that all hadrons are actually bound states of fundamental spin 1/2 particles called quarks. Whereas all other known particles have an electric charge equal to where is the proton charge, quarks have electric charges equal to $\pm e$ either $-e/3$ or $+2e/3$. Leptons themselves are considered to be fundamental, so the leptons and the quarks form the basic building blocks of all matter in the universe.

Quarks are subject to electromagnetic forces via their charge, but interact most strongly via the so-called strong force. The strong force is carried by massless, uncharged, spin 1 bosons called gluons.

Table. Table of Quark Types, Charge, Rest Energy (in GeV), and the Four "Exotic" Flavour Quantum Numbers.

Type	Charge	Rest energy	s	c	b	t
down (d)	–1/3	0.333	0	0	0	0
up (u)	+2/3	0.330	0	0	0	0
strange (s)	–1/3	0.486	–1	0	0	0
charm (c)	+2/3	1.65	0	+1	0	0
bottom (b)	–1/3	4.5	0	0	–1	0
top (t)	+2/3	176	0	0	0	+1

When Murray Gell-Mann and George Zweig first proposed the quark model in 1963, they needed to postulate only three types or flavours of quarks, up, down, and strange. These were sufficient to explain the constitution of all hadrons known at the time. We currently know of six different flavours of quarks. Their properties are listed in table. The properties charm, topness, and bottomness are analogous to strangeness these properties are conserved in strong interactions. Weak interactions, can turn quarks of one flavour into another flavour. However, the strong and electromagnetic forces cannot do this.

Just as the proton and the neutron have antiparticles, so do quarks. Antiquarks of a particular type have strong and electromagnetic charges of the sign opposite to the corresponding quarks. Quarks have baryon number equal to 1/3, while antiquarks have –1/3. Thus combining three quarks results in a baryon number equal to 1, while together a quark plus an antiquark have baryon number zero. All baryons are thus combinations of three quarks, while all mesons are combinations of a quark and an antiquark. Table lists a sampling of hadrons and some of their properties. Notice that the same combination of quarks can make up more than one particle, *e.g.*, the positive pion and the positive rho. The positive rho may be considered as an excited state of the u *d* system, while the positive pion is the ground state of this system.

Table. Sample Hadrons

Type	Charge	Rest energy	Spin	Composition	Mean life
Proton	+1	938.280	1/2	uud	stable
Neutron	0	939.573	1/2	udd	898
Lambda	0	1116	1/2	uds	3.8 × 10–9
positive pion	+1	140	0	u *d*	2.6 × 10–8
negative pion	–1	140	0	*u* d	2.6 × 10–8
neutral pion	0	135	0	u *u* - d *d*	8.7 × 10–17
positive rho	+1	770	1	u *d*	4 × 10–24
positive kaon	+1	494	0	u *s*	1.24×10–8
neutral kaon	0	498	0	d *s*	8.6 × 10–11
J/psi	0	3097	1	c *c*	1.5 × 10–20

Yet to be mentioned is the quantum number colour, which has nothing to do with real colours, but has analogous properties. Each flavour of quark can take on three possible colour values, conventionally called red, green, and blue. Antiquarks can be thought of as having the colours antired, antigreen, and antiblue, also known as cyan, magenta, and yellow. Because of this, the theory of quarks and gluons is called quantum chromodynamics. Counting all colour and flavour combinations, there are $6 \times 3 = 18$ quarks.

As in electromagnetism, the strong force has associated with it a "strong charge". However, this charge is some what more complicated than electromagnetic charge, in that there are three kinds of strong charge, one for each of the strong force colours. Each colour of charge can take on positive and negative values equal to $\pm gs$. As with electromagnetism, positive and negative charges (of the same colour) cancel each other. However, in quantum chromodynamics there is an additional way in which charges can cancel. A combination of equal amounts of red, green, and blue charges results in zero net strong charge as well.

Gluons, the intermediary particles of the strong interaction come in eight different varieties, associated with differing colour-anticolour combinations. Since gluons don't interact via the weak force, there is no flavour quantum number for gluons — quarks of all flavours interact equally with all gluons.

The quark model of matter has led to extensive searches for free quark particles. However, these searches for free quarks have proven unsuccessful. The current interpretation of this result is that quarks cannot exist in a free state, basically because the attractive potential energy between quarks increases linearly with separation. This appears to be related to the fact that gluons, the intermediary particles for the strong force, can interact with each other as well as with quarks. This leads to a series of increasingly complex processes as quarks move farther and farther apart. The result is called quark confinement — apparently, individual quarks can never be observed outside of the confines of the observable particles which contain them. Confinement works not only on single quarks, but on any "coloured" combinations of quarks and gluons, *e.g.*, a red up quark combined with a green down quark. It appears that long range inter-quark forces only vanish for interactions between "white" or "colour-neutral" combinations of quarks. This is why only colour-neutral combinations of quarks — three quarks of three different colours or a quark-antiquark pair of the same colour — are actually seen as observable particles.

The strong equivalent of the fine structure constant is the coupling constant for the strong force:

$$a_s = \frac{g_s^2}{4\pi\hbar c}$$

However, they are still relatively strong compared to, say, electromagnetic forces. These residual strong forces are responsible for nuclear processes.

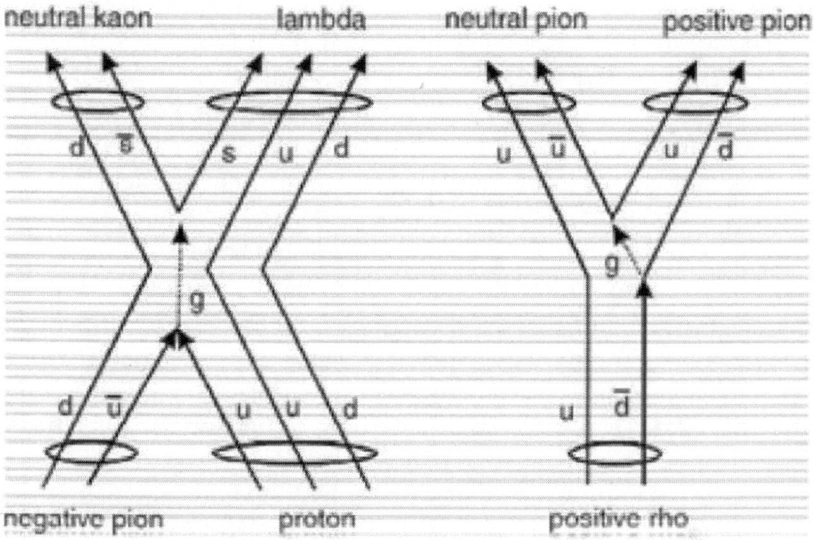

Fig. Some Sample Strong Interactions Illustrated in Terms of Gluon

Emission and Absorption

Interactions between hadrons can be thought of as resulting from interactions between the individual quarks making up the hadrons. Two sample strong interactions. Virtual gluons can be emitted and absorbed by quarks much as virtual photons can be emitted and absorbed by electrically charged particles. Particles unstable to strong decay processes (such as the positive rho particle) typically live only about10^{-23} $_s$, where as particles stable to strong decay but unstable to weak decay live of order $10^{-10}s$ or longer, depending strongly on how much energy is liberated in the decay. Particles subject to electromagnetic decay processes, such as the neutral pion, take on mean lifetimes intermediate between strong and weak values, typically of order $10^{-18}\,s$.

QUANTUM THEORY

We have traversed the realms of physical research as rapid travellers. Since we failed to take along the heavy weapons of mathematics, we have been limited, so to speak, to the role of a civilian war correspondent in the land of physical research who must be content to draw some mood pictures of the events there without seriously pursuing the strategical and technical problems. We passed by rich departments of knowledge, varied and beautiful in their content.

To omit consideration of the thought structures in which modern physical thought triumphs over the apparently hopeless paradoxes and difficulties which were evidenced in quantum physical experiments would require still greater decision. It is not possible to relate and explain here the heavy weight of experimental quantitative proof of the content and mathematical laws of modern quantum theory;

a clarification of the philosophical-logical character of this theory by intimation must suffice. But even for that the author, who wants to make it more easily accessible to the reader, must request special attention.

Let us consider briefly the historical development of modern quantum theory. The investigation of atomic spectra (and the energy levels related to them) presented physical research with an abundance of varied problems, in the step by step clarification of which physicists gradually worked out an understanding of the regularities of quantum physics. Niels Bohr, who started this development in 1913 also remained its leader.

Without going into the thousand different problems of this field we shall consider the Bohr "correspondence principle". Although it was mentioned in reference to the energy or relativity principle — and not opposing these in its meaning — the correspondence principle has a quite different character from the energy and relativity principles. These latter are natural laws in finished, perfected form; their content can be expressed clearly and thoroughly in a few words. But the correspondence principle presents peculiar difficulties to the intelligence since its general content can only be described by intimation, in special individual examples. For the correspondence principle is not a finished, clearly definable law of nature, but indicates the direction in which Bohr's conviction about the solution of the quantum puzzle was to be sought, and in which he wished to steer the reflection of the quantum investigators.

It only becomes clear in the history of quantum theory that not only individual great discoveries are decisive for the development of our knowledge but that under certain circumstances the change of our spiritual attitude towards problems can be much more important. Physical thought develops not only from combining the compilations of results of individual observations and their logically-exact treatment; but rather the attainment of decisive new intelligence depends essentially on a creative phantasy, which for its part depends essentially on spiritual hypotheses and on the attitude our intuition assumes towards these things.

Bohr's historical contribution lies not only in the individual, pioneer discoveries through which he enriched quantum theory but also in the penetrating force his spirit exerted for the creation of a new spiritual "atmosphere" — wherein the problems were first gradually elucidated as the essential became distinguishable from the unessential and effective control of the solution became possible. The meaning of the correspondence principle is not purely historical; it is still indispensable today in the sense that it teaches us to view modern quantum theory knowledge with the proper attitude.

The differences between quantum theory and classical physics are so deeply rooted that many physicists were tempted to regard the ideas of classical physics as absolutely useless for the comprehension of atomic physics; they attempted to introduce arbitrarily invented, new, untraditional ideas. But Bohr had — and this is the content of the correspondence principle — nergetically pointed out that in all individual problems a verifiable close similarity exists between classical and quantum theory despite their fundamental difference.

The following is a rough example: classical physics predicts that atoms, which con sist of electrically charged particles, necessarily emit light in the execution of internal motions. This prediction is correct. The elementary processes of this light emission do proceed quite differently from macroscopic antennae; but the fact remains that the fundamental expectation is fulfilled. That is an example of the close relationship which exists between classical and quantum theory despite their incisive difference. There are other examples of this relationship, much finer examples; and through Bohr we gradually learned to see that such kindred relationships can absolutely be uncovered for each concrete individual problem through more exact analysis.

In his emphasis on the affinity of classical and quantum theory there exists, however, a decided prominence of the independence of the quantum theory from classical theory. Since we learn to "understand" various quantum- physical individual problems better through uncovering kindred relationships "in the manner of correspondence", we gradually attain one of the concepts of quantum physics used in classical theory but separated from it by fundamental differences and independent in itself. Such a discovery is a radical challenge to the repeated spasmodic attempts throughout the course of historical development to somehow want to "explain" the characteristic quantum phenomena — to want to reduce them to ideas which conform more closely to the classical.

With the progress of this development it gradually became clear which problems in general should be answered by a logically effected quantum theory. Here the positivist consideration must be repeated very forcefully; the goal of a purely intellectual comprehension of quantum phenomena must be a description of the experimental facts themselves. The experimental facts involv ed h ere a re collectively of the follow ing sort: a spec ific atom, characterized by its atomic number, according to experience possesses quite definite energy levels. The first problem appears; how can the positions of all these energy levels (for any atom, and likewise for molecules) be determined and calculated from general theoretical principles? Furthermore, we find "transition-probabilities" for these atoms. When an atom exists in an energy- rich condition, after some time it will undergo a quantum transition which brings it down to a lower energy level while the energy freed thereby comes off as a light quantum.

The atom has a range of various such possibilities since there are different lower energy levels at its disposal. It is not possible to predict in an individual case when and where the atom will jump. But experience indicates the existence of quite definite probabilities for the various possible processes. If the atom is irradiated with light there results a definite probability for a quantum transition associated with the absorption of a light quantum. If an electron of a certain (sufficiently high) velocity impinges on an atom, the atom can, as we already know, thus also be induced to a quantum transition. If, furthermore, the atom collides with another atom (of the same or a different kind) an energy change can occur, as was already mentioned, in the form of a simultaneous mutual quantum transition.

Definite probabilities prevail for all these processes; based on experience they are always stated in the same ways through the conditions of the experiment in question. They are summarized under the designation of "tra ns ition proba bilities" . Now the problem (b esides the theoretic al determination of energy levels) that a complete quantum theory must solve quite comprehensively is simply the theoretical determination of the transition probabilities. In this sense Heisenberg undertook the creation of a "quantum mechanics". He relied upon the success then already achieved in the detailed execution of Bohr's correspondence principle; in fact, the systematic evaluation of Heisenberg's extensions (Heisenberg, Born, Dirac, Jordan) yielded a solution of the problem formulated according to the correspondence principle which was complete in principle. Of course the mathematical formulation of this solution was such that it differed extensively from the mathematically precise renditions of our classical physical conceptions. But this mathematical form of the new quantum mechanics was most exactly suited to the problem (as we stated it) of the determination of transition probabilities. Besides, this new quantum mechanics expressed wonderfully the two-sided relation of quantum to classical theory; namely, fundamental difference on the one side and close connection on the other.

These investigations had just reached a preliminary rounding-off point when Schrödinger arrived at the same goal by an entirely different method. Schrödinger started with de Broglie's investigations. After de Broglie had shown how the uniform rectilinear motion of a corpuscular particle was reinterpreted in the wave theory, Schrödinger wondered how these de Broglie considerations developed for motions influenced by external forces. For this reason he investigated the motion of an electron under the influence of the attraction of a positive heavy nucleus; and thus he arrived at the quantitative description of the electron charge cloud in the hydrogen atom.

Schrödinger could now show that with the solution of the mathematical problem he formulated he had simultaneously also achieved the solution for the apparently quite different mathematical problem which was expressed by quantum mechanics in the form of matrix theory. Thus, when a problem has been solved by means of Schrödinger's "wave mechanics" a mathematical conversion yields the "quantum mechanics" solution for this same problem. This mathematical connection between the two theories, leading to the same result, which in view of the complete difference of the two methods must appear very surprising at first, also supplied the certain basis for the physical- abstract interpretation of Schrödinger waves.

At first it might appear that the discovery of wave mechanics yielded a mitigation of the contrast between quantum and classical theory; for here instead of the unclassical concepts of transition probabilities, etc., we deal with waves — something closer to classical ideas. But what we just learned with respect to light is still true; the classical wave theory is not the final word.

Wave mechanics does not in any way signify for atomic physics the removal or mitigation of the fundamental unclassical characteristics of quantum physics. Also, for atoms with more than one electron (thus for all atoms except hydrogen)

Schrödinger wave mechanics assumes a very abstract form; in these cases Schrödinger waves are no longer waves in customary three-dimensional space but are simply a mathematical construction which mathematicians can "illustrate" to themselves as waves in space of more than three dimensions.

This abstract, multidimensional space can be avoided through another (mathematically equivalent) method of representation, the construction of which was a special hobby of the author's. In this method of representation ("second quantization"), which clings especially closely to the fundamental dualism of waves and corpuscles, the waves dealt with are spread out in ordinary three-dimensional space, but can only be described by means of the ideas of quantum mechanics.

These theories, whose abstract nature will not remain hidden from the reader in even this fleeting explanation, made possible a number of special uses for individual problems of atomic physics. The precise correctness of the new theory was confirmed without exception in a tremendous field of experimental results. No end is in sight for the further elaboration of special problems on the basis of these principles. The fundamental result may be stated as follows: today we can understand all phenomena which occur near and in the electron shells. Thus all the elementary processes that are the basis for spectral or chemical processes are defined as clearly as the motions of the planetary systems have been since Newton. The only partially explored realm at present remains the physics of internal events in atomic nuclei. For us it is only important to understand the philosophical nuclei of these new ideas. We again owe special thanks to Bohr and Heisenberg for the philosophical-epistemological explanation of the meaning and significance of the new theories. This became possible through the earlier thorough mathematical understanding of the quantum laws in the so-called "statistical transformation theory" (Dirac, Jordan).

ORIGIN OF THE QUANTUM THEORY

The quantum theory had its origin in an effort to account for the radiation by black bodies. A *black body* is generally understood to be a body which absorbs all the light falling upon it and, accordingly, can reflect none. Physicists, on the other hand, use the term in a much broader sense; by their interpretation a perfect black body is one which at all temperatures absorbs all the radiation falling upon it, whether it be visible light, ultraviolet, or infrared radiation. Moreover the black body is an ideal radiator: the energy emitted per unit area at any temperature is equal to that absorbed. However, the nature of the radiation emitted by a black body depends on its temperature.

The emitted radiation is not uniform but has a maximum value at a definite wavelength which is inversely proportional to the absolute temperature, and the total energy that is emitted varies as the fourth power of the temperature. A heated solid emits radiation through a whole range of wavelengths, and if the energy is plotted in the form of a graph against the wavelength, the curve shows a maximum.

The classical theory predicted that most of the energy should be in the form of the shortest wavelengths. Furthermore, according to the classical theory, the light emitted by a body should remain the same colour, only its intensity altering with temperature. It is a well-known fact that a heated body changes through red and yellow to white as its temperature increases.

To overcome the difficulties in explaining the black-body spectrum, Max Planck, in 1900, arrived at the revolutionary conclusion that radiation is emitted and absorbed discontinuously in integral multiples of a fundamental unit which he designated as a quantum, the energy of this unit being proportional to the frequency of light. This conclusion led to the development of the quantum theory.

Planck's Quantum Theory

A black body is one which will absorb completely incident radiation of any wavelength. The ideal black body does not reflect any energy, but it does radiate; in fact it is the most efficient possible emitter of radiation. For this reason such a body is sometimes called a *complete radiator*. Although no actual

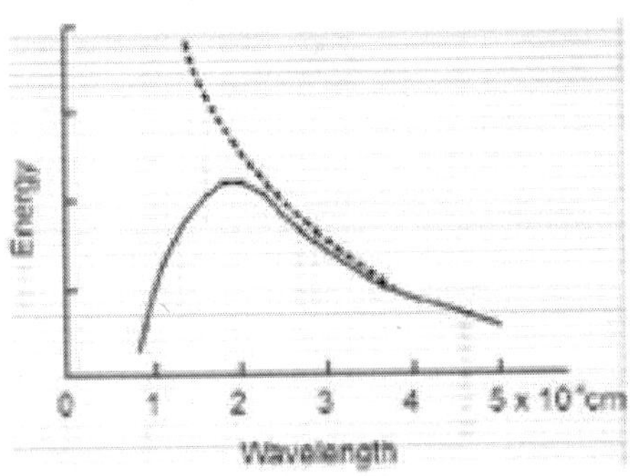

Fig. Spectral Distribution of Energy of Black-body Radiation at 1500°K. The Curve Represents the Observed Distribution of Energy Among Different Wavelengths. The Dotted Line Represents the Distribution of Energy as Predicted by the Classical Theory.

body is perfectly black, it is possible to postulate a condition where radiation is completely enclosed in a space surrounded by thick walls at a constant temperature. Thus we may imagine a cavity inside a solid body containing a small hole in one wall whereby the investigator may study the enclosed radiation without altering its character.

The light emerging through the hole has a distinctive spectrum, usually referred to as the *black-body spectrum*, and, in accordance with our original assumptions, it will be identical with the spectrum of the light inside the cavity. If the energy emitted is plotted against its frequency or wavelength, where $v = c/L$, the curve shows a maximum. This graph shows that the energy emittedis greatest at the middle wavelengths in the spectrum and least at the highest and lowest frequencies. If curves are plotted for a series of temperatures, as given in figure it is found that the maximum moves towards shorter wavelengths as the temperature rises.

Max Planck, in 1900, developed an empirical equation which satisfactorily represented the experimental data obtained for the energy distribution in the spectrum of a black body. However, he was unable to justify this equation by theoretical reasoning upon the basis of the classical electromagnetic wave theory. Furthermore the classical conceptions of continuity completely broke down in an endeavor to explain the results of these experiments.

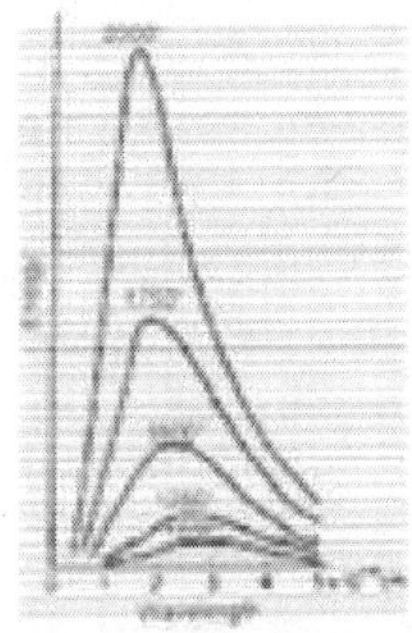

Fig. Energy Distribution for Blackbody adiation at Different Absolute Temperatures.

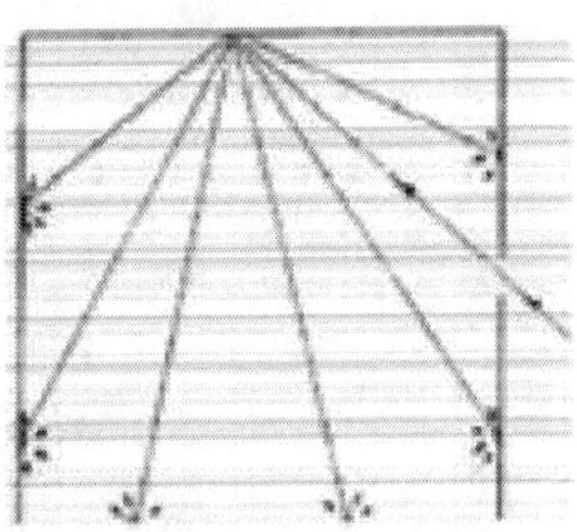

Fig. A Black-Body Enclosure with a Small Opening in One Wall.

Consequently Planck reached the conclusion that the laws of classical physics do not apply to energy transformations on an atomic scale. In order to explain the black-body spectrum, he postulated that all solid bodies, including the walls of the cavity, consisted of an enormous number of elementary oscillators, vibrating at all possible frequencies. These oscillators, which he called "resonators," were conceived by Planck to be small electrified particles in constant motion; later he identified them as vibrating electrons, but in 1900 the electronic concept of matter was not well established.

According to this theory there must be oscillators in all radiating bodies for any possible range of frequencies. Moreover the energy of any one oscillator of frequency v could assume only such values as were integral multiples of a unit e, which in turn was proportional to the frequency v, the proportionality factor being the universal constant h. This postulate was expressed in the equation.

$$e = hv$$

Where h is usually called Planck's constant. In other words, the energy of any one oscillator was not permitted to take any arbitrary value but was proportional to some integral multiple of the frequency of the oscillator. This was a radical break with the classical theory, which postulated that all such oscillators must have the same energy, no matter what their frequency, and that, although the amount of energy for any oscillator might vary, the variation must be a continuous one, up or down, through the energy scale.

In explaining the experimental data for black-body radiation, as shown in figure at low temperatures, only those oscillators with low natural frequencies are excited; therefore the radiating body can emit only light of low frequency. Since more oscillators have high frequencies, more radiation takes place as the temperature is raised. Accordingly the energy of the black body is not continuously variable but is emitted in small packets, and the number of packets is increased with temperature.

Photoelectric Effect

From the study of X-ray spectra we have learned that, when particles of matter with suitable velocities strike a metal surface, the energies of the particles can be absorbed by the surface and then emitted in the form of radiation. The reverse transformation is also possible. A beam of light of suitable frequency falling upon a metal surface will cause the emission of electrons. This phenomenon is called the *photoelectric effect* and provides the basis for the photoelectric cells used for commercial purposes. Although this effect was first observed by Heinrich Hertz in 1887, credit is usually given to Hallwachs, who, in 1888, announced that a freshly polished zinc plate would lose its charge when charged negatively and illuminated by ultraviolet light but would not lose its charge when charged positively.

The spectral sensitivity of a surface depends upon the nature of the metal used. Those metals which are the best conductors of electricity, and presumably have the largest number of' free electrons, are not the ones that show the greatest photoelectric action. There seems to be some relation between chemical reactivity and photoelectric effect, since those metals which are most active, photoelectrically, are those with few electrons in their outermost shells, particularly those which have only one, such as sodium, potassium, etc.

Thus the feeble light of a candle falling upon the surface of the alkali metals is sufficient to cause the emission of electrons. Those metals with two electrons in their outermost shell, *e.g.*, magnesium and zinc, act in a similar manner when subjected to radiation of greater frequency, such as ultraviolet light. Other metals are also slightly active, although X rays and gamma rays are sufficiently powerful to cause the emission of electrons from all forms of matter. These facts lead us to conclude that the energy of the electron which is emitted comes from the light beam, and not from the atom itself. Electrons emitted by the action of radiant energy are called *photoelectrons*. The maximum velocity attained by these electrons varies directly with the frequency but is independent of the intensity of' the radiation that liberates them. A very weak beam of light of high frequency can liberate electrons that travel with a high velocity, whereas a low-frequency ray of light, regardless of intensity, may liberate only lowspeed electrons. An increase in intensity merely serves to increase the number of photoelectrons.

Einstein's Concept of Light Quanta

In 1905 Einstein extended the quantum theory, as proposed by Planck, by suggesting that radiation is not only absorbed and emitted in quanta but that it is propagated through space in definite quanta called *photons*. The energy of each individual light quantum, or photon, is given by the expression hv, where h is Planck's constant and v is the frequency of the light. Einstein also suggested that the photoelectric process could be explained by assuming that the energy of a photon, hv, was absorbed by an electron and converted into kinetic energy, whereby the electron was enabled to escape from the surface of a metal. The equation he developed for the process was

$$1/2mv2 = hv - W$$

in which the expression $1/2mv2$ denotes the kinetic energy of the emitted electrons, hv is the energy of a photon, and W is the work required to remove the electron from the metal surface. Thus the kinetic energy of the emitted electron will always be less than that of the incident photon because of the energy necessary to release the electron from the surface. The equation also explains why the number of emitted electrons is proportional to the intensity of the light.

The foregoing mathematical expression is known as *Einstein's photoelectric equation*. The validity of the equation has been justified by numerous experiments, and it also has served as a basis for determining the value of Planck's constant h. The interpretation of the equation has been simplified by the modern theory of metals. A metallic conductor is believed to be made up of a lattice of positive ions, permeated by loose electrons which are in random motion. An electron may come to the surface of a metal, but it cannot escape because of a potential difference between the metal and the surrounding space.

The photoelectric effect is produced by light of sufficiently short wavelength falling upon a metal surface and imparting to some of the electrons enough energy to permit them to penetrate the potential barrier. However, unless the frequency of the incident radiation exceeds a minimum value, so that hv is greater than W, there can be no photoelectric emission of electrons. It is an all-or-none process in which the electron either acquires all the energy of the photon and escapes from the surface or accepts none of it and remains within the potential barrier of the metal.

The Compton Effect

Additional evidence of the corpuscular nature of electromagnetic radiation (light) is furnished by the change in wavelength of X rays when they are scattered upon impact with atoms or molecules. This phenomenon was observed by A. H. Comptonin 1923 and is therefore usually referred to as the *Compton effect*. It differs from the photoelectric effect in respect to the amount of energy absorbed by an electron within a surface. The photoelectric effect involves the complete absorption of a quantum, whereas in the Compton effect only a portion of a quantum is imparted to an electron.

According to the classical electromagnetic wave theory, monochromatic light falling upon matter should be scattered without change in frequency. However, when Compton permitted monochromatic X rays to impinge upon matter of low atomic weight, he observed that secondary X rays of slightly longer wavelength than those of the impinging beam were produced.

In other words, when a photon collides with an electron, it may leave with a lower frequency, and the electron thereby acquires a greater velocity. In the collision the energy of the photon is reduced, and the energy of the electron is increased.

The Compton effect may be explained as resulting from an impact between two bodies, the photon and the electron, which is, of course, additional proof of the corpuscular nature of light. This impact is elastic, and both kinetic energy and momentum are conserved. After the encounter each body has a different energy and a different momentum from the values it had before the contact, but the sum of the energies and the sum of the momenta are unchanged.

Compton assumed that the same principles for conservation of energy and for change in momentum apply to the impact between a photon and an electron as apply to the collision of any other two moving bodies. The energy of a quantum is equivalent to *hv*, and if the quantum is considered as a particle, which we shall call a photon, it possesses momentum equivalent to *hv/c,* where c is the velocity of light. When the photon collides with an electron, the electron gains momentum and the photon is deflected with a loss of momentum.

Since h and c are constants and do not change, the energy transformation is manifested as a lowering of frequency (wavelength is increased) for the photon as a result of the collision. Following the encounter between the two bodies, the photon moves in one direction at an angleq with momentum equal to *hv*Q/*c,* and the electron recoils in another direction at an angle ¢ , with momentum equal to where b is the ratio of the velocity of the electron to the velocity of light. It can be shown mathematically that

$$^{mv}\sqrt{1-\beta^{2}},$$

where β is the ratio of the velocity of the electron to the velocity of light. It can be shown mathematically that

Fig. Collision Between a Photon and an Electron; the Compton Effect.

the change in wavelength, where λ c/p, of the photon is given in the equation

$$\lambda = \frac{h}{mc}(1 - \cos{}'A)$$

Where m is the rest mass of the electron. In asmuch as *h, m,* and *c* are all constants, the equation can be simplified to

$$\lambda = 0.024(1 - \cos{}'A)$$

The evidence of the Compton effect is of great significance. It indicates that X rays normally thought of as radiation can also be treated as particles, using classical laws.

OLD QUANTUM THEORY

Quantum mechanics developed from the study of electromagnetic waves through spectroscopy which includes visible light seen in the colours of the rainbow, but also other waves including the more energetic waves like ultraviolet light, x-rays, and gamma rays; and the waves with longer wavelengths including infrared waves, microwaves and radio waves. Only waves that travel at the speed of light are included in this description. Also, when the word "particle" is used below, it always refers to elementary or subatomic particles.

PLANCK'S CONSTANT

Classical physics predicted that a black-body radiator would produce infinite energy, but that result was not observed in the laboratory. If black- body radiation was dispersed into a spectrum, then the amount of energy radiated at various frequencies rose from zero at one end, peaked at a frequency related to the temperature of the radiating object, and then fell back to zero. In 1900, Max Planck developed an empirical equation that could account for the observed energy curves, but he could not harmonize it with classical theory. He concluded that the classical laws of physics do not apply on the atomic scale as had earlier been assumed.

In this theoretical account, Planck allowed all possible frequencies, all possible wavelengths. However, he restricted the energy that is delivered. "In classical physics,... the energy of a given oscillator depends merely on its amplitude, and this amplitude is subject to no restriction." But, according to Planck's theory, the energy emitted by an oscillator is strictly proportional to its frequency.

The higher the frequency, the greater the energy. To reach this theoretical conclusion, he postulated that a radiating body consisted of an enormous number of elementary oscillators, some vibrating at one frequency and some at another, with all frequencies from zero to infinity being represented. The energy *E* of any one oscillator was not permitted to take on any arbitrary value, but was proportional to some integral multiple of the frequency *f* of the oscillator. That is,

$$E = nhf.$$

where $n = 1, 2, 3,...$ The proportionality constant h is called *Planck's constant.* One of the most direct applications is finding the energy of photons. If *h* is known, and the frequency of the photon is known, then the energy of the photons can be calculated. For instance, if a beam of light illuminated a target, and its frequency was 540 × 1012 hertz, then the energy of each photon would be h × 540 × 1012 joules. The value of h itself is exceedingly small, about 6.6260693 × 10-34 joule seconds. This means that the photons in the beam of light have an energy of about 3.58 × 10-19 joules or approximately 2.23 eV.

When the energy of a wave is described in this manner, it seems that the wave is carrying its energy in little packets. This discovery then seemed to remake the wave into a particle. These packets of energy carried along with the wave were called quanta by Planck. Quantum mechanics began with the discovery that energy is delivered in packets whose size is related to the frequencies of all electromagnetic waves (and to the colour of visible light since in that case frequency determines colour). Be aware, however, that these descriptions in terms of packet, wave and particle import macro-world concepts into the quantum world, where they have only provisional relevance or appropriateness.

In early research on light, there were two competing ways to describe light, either as a wave propagated through empty space, or as small particles travelling in straight lines. Because Planck showed that the energy of the wave is made up of packets, the particle analogy became favoured to help understand how light delivers energy in multiples of certain set values designated as quanta of energy. Nevertheless, the wave analogy is also indispensable for helping to understand other light phenomena.

In 1905, Albert Einstein used Planck's constant to explain the photoelectric effect by postulating that the energy in a beam of light occurs in concentrations that he called light quanta, later on called photons. According to that account, a single photon of a given frequency delivers an invariant amount of energy.

In other words, individual photons can deliver more or less energy, but only depending on their frequencies. Although the description that stemmed from Planck's research sounds like Newton's corpuscular account, Einstein's photon was still said to have a frequency, and the energy of the photon was accounted proportional to that frequency. The particle account had been compromised once again.

Both the idea of a wave and the idea of a particle are models derived from our everyday experience. We can only investigate their properties indirectly. We look at some phenomena, such as the rainbow of colours that we see when a thin film of oil rests on the surface of a puddle of water, and we can explain that phenomenon to ourselves by comparing light with waves. We look at other phenomena, such as the way a photoelectric meter in our camera works, and we explain it by analogy to particles colliding with the detection screen in the meter. In both cases we take concepts from our everyday experience and apply them to a world we have never seen.

Neither form of explanation, wave or particle, is entirely satisfactory. In general any model can only approximate that which it models. A model is useful only within the range of conditions where it is able to predict the real thing with accuracy.

Newtonian physics is still a good predictor of many of the phenomena in our everyday life. To remind us that both "wave" and "particle" are concepts imported from our macro world to explain the world of atomic-scale phenomena, some physicists such as Banesh Hoffmann have used the term "wavicle" to refer to whatever it is that is really there.

REDUCED PLANCK'S CONSTANT

Planck's constant originally represented the energy that a light wave carries as a function of its frequency. A step in the development of this concept appeared in Bohr's work. Bohr was using a "planetary" or particle model of the electron, and could not understand why a 2π factor was essential to his experimentally derived formulae. Later, de Broglie postulated that electrons have frequencies, just as do photons, and that the frequency of an electron must conform to the conditions for a standing wave that can exist in a certain orbit.

That is to say, the beginning of one cycle of a wave at some point on the circumference of a circle must coincide with the end of some cycle. There can be no gap, no length along the circumference that is not participating in the vibration, and there can be no overlap of cycles. So the circumference of the orbit, C, must equal the wavelength, λ, of the electron multiplied by some positive integer ($n = 1, 2, 3...$). Knowing the circumference one can calculate wavelengths that fit that orbit, and knowing the radius, r, of the orbit one can calculate its circumference. To put all that in mathematical form, and so,

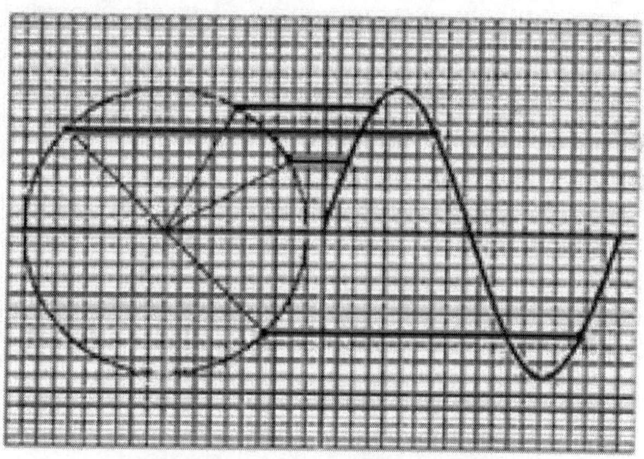

Fig. Relation between a Cycle and a Wave

$$C = 2\pi r = n\lambda$$

and the appearance of the 2p factor is seen to occur simply because it is needed to calculate possible wavelengths (and therefore possible frequencies) when the radius of an orbit is already known.

Again in 1925 when Werner Heisenberg developed his full quantum theory, calculations involving wave analysis called Fourier series were fundamental, and so the "reduced" version of Planck's constant (*h*/2p) became invaluable because it includes a conversion factor to facilitate calculations involving wave analysis. Finally, when this reduced Planck's constant appeared naturally in Dirac's equation it was then given an alternate designation, "Dirac's constant." Therefore, it is appropriate to begin with an explanation of what this constant is, even though the theories that made its use convenient.

The energy of any wave is given by its frequency multiplied by Planck's constant. A wave is made up of crests and troughs. In a wave, a cycle is defined by the return from a certain position to the same position such as from the top of one crest to the next crest. A cycle actually is mathematically related to a circle, and both have 360 degrees. A degree is a unit of measure for the amount of turn needed to produce an arc of a certain length at a given distance. A sine curve is generated by a point on the circumference of a circle as that circle rotates.

There are 2 π radians per cycle in a wave, which is mathematically related to the way a circle has 360° (which are equal to two π radians). (A radian is simply the angle if a distance along the circumference of the circle is measured equal to the radius of the circle, and then lines are drawn to the centre of the circle. This forms an angle equal to 1 radian.) Since one cycle is 2 π radians, when h is divided by 2 π the two "2 π " factors will cancel out leaving just the radian to contend with. So, dividing h by 2 π describes a constant that, when multiplied by the frequency of a wave, gives the energy in joules per radian. The reduced Planck's constant is written in mathematical formulae as ",and is read as "h-bar".

$$\hbar = \frac{h}{2\pi}$$

The reduced Planck's constant allows computation of the energy of a wave in units per radian instead of in units per cycle. These two constants *h* and ' are merely conversion factors between energy units and frequency units. The reduced Planck's constant is used more often than *h* (Planck's constant) alone in quantum mechanical mathematical formulas for many reasons, one of which is that angular velocity or angular frequency is ordinarily measured in radians per second so using ' that works in radians too will save a computation to put radians into degrees or vice-versa. Also, when equations relevant to those problems are written in terms of', the frequently occurring 2 π factors in numerator and denominator can cancel out, saving a computation. However, in other cases, as in the orbits of the Bohr atom, *h*/ 2 π was obtained naturally for the angular momentum of the orbits. Another expression for the relation between energy and wave length is given in electron volts for energy and angstroms for wavelength: $_{Ephoton}$ $(\text{eV}) = 12{,}400 / \lambda(\dot{A})$— it appears not to involve *h* at all, but that is only because a different system of units has been used and now, numerically, the appropriate conversion factor is 12,400.

CLASSICAL THEORY

We return once again to classical theory to emphasize still more clearly the characteristic features through which it differs from quantum theory. We can state three basic principles which should be considered as the most important characteristics of classical theory with the catchwords continuity, causality, objectivity. We spoke amply of continuity; and we learned that just this principle of continuity constitutes a difference between classical and quantum theory. But now we want to make it clear that continuity is not only something familiar to us, but is really an essential, the omission of which is followed with logical necessity by further fundamental deviations from classical ideas.

The assumed continuity of natural processes is essential for our method of executing and evaluating physical measurements. Every measurement we perform is inexact. The physicist considers the "limit of error" for each measurement; for every measured numerical value he records two numbers, the differenc e between them corresponding to the uncertainty of the measurement. That it is nevertheless possible to draw certain conclusions from an inexact measurement — conclusions, of course, that are also affected by an uncertainty but in any case are rich in content and very definite — is only because the principle of continuity is actually fulfilled in the macrophysical world; trivial changes in the cause are followed by trivial changes in the effect. Consequently an inexact knowledge of causes is still sufficient for an inexact prediction of the effects, although the absolutely certain prediction of the effects becomes possible only with a mathematically more precise (practically unattainable) measurement of the causes.

Uncertainty in the process of measurement deserves a still more detailed investigation. The essential point for us here will become clear if we visualize, for example, the measurement of the temperature of a macrophysical body. We bring the body in contact with a thermometer; the thermometer assumes the same temperature as the body in question; and we read off the value. But to be exact, it is necessary to consider that since the body under investigation imparts some of its heat to the thermometer the body itself is influenced and its original condition is altered somewhat.

To avoid resultant false measurements the thermometer selected must be much smaller than the body being investigated. (Of course it is possible, if the thermometer used was not sufficiently small, to compute or estimate the temperature change which took place and correct the result for it; but that is a problem in technical method which has no connection with the nature of the thing.) It is analogous for every physical measurement: I must always select such fine instruments for a measurement that in the measuring process the body under investigation itself will not be influenced appreciably by too rough a measuring instrument (such influence would falsify the measured result).

Fundamentally a reaction of the measuring instrument on the object being investigated is inherent in every physical measurement; yet in the investigation of macrophysical objects this reaction can be made sufficiently small by the selection of quite fine instruments.

This idealization of the process of measuring in classical physics depends essentially on the assumption of continuity of all natural processes in the sense that our fundamental considerations are established as though precise measurements were possible for us. It is a "permissible idealization" to base our considerations on this concept of absolute observational accuracy and to interpret the actual inaccuracy of each measurement as an only practically and not principally significant secondary disturbance. This idea cannot lead us to errors, because by classical concepts we can approach ideal measurement unlimitedly, although we can never attain it.

We here consider something which we shall indicate with the word objectivity and which is so characteristic of our general classical-physical thinking that we usually disregard it. We are accustomed to regarding physical observation results as clearly understood according to their importance if we have explained them as effects of an objective physical process or condition.

This formulation hides within it a higher type of positivist modesty; a deeper "explanation" of natural processes is relinquished herein for everything. But what is left after this renunciation of all classical-physical theories as a basis for their methods of representation is just this idea in objective events. Perhaps we do not describe planetary motions by specifying when and where or through what telescope the various planets were observed; but we do describe planetary motions as a spatial-temporal process taking place independently of human observation.

Naturally one can pursue the familiar philosophical considerations which emphasize that without subjects doing the observing there can be no talk of objective matters. But these thought processes are not acceptable to the physicist. To him the motion of the planet Neptune in its orbit is an objective event which was already in process before anyone had seen this planet in a telescope and which continues uninfluenced, independent of whether or when or how often it is observed or photographed. Exactly the same self-evident supposition of objective events is the basis for Maxwellian electrodynamics; we interpret electrical measurement data as indication of an objective physical event in the electromagnetic field which is present in space.

Positivist criticism must remind us that this objectivity of physical events is not a purely logical self-evident fact. For positivism teaches us to view true physical reality only in the totality of experimental results. It is very remarkable and astonishing that in the domain of validity of macrophysics we are in a position to so formulate our summarizing description of experimental results that they are no longer referred to directly but are, so to speak, only minor appendages of the picture we have traced out — a picture which maintains the existence of objective events which occur independently of how and where the observations necessary for their detection are taken.

After what has been said it is obvious that the idea of ideal measurements (made possible by the continuity of macrophysical processes) which do not disturb the observed event in the least is in dispensable for this construction of an objective physical world. Another assumption indispensable for this objectivity is that of complete causality in the macrophysical world. For if we had to reckon with the occurrences of effects not determined strictly causally we could never have been certain in our observation processes whether the source of an effect we saw was to be sought in the object we were observing or just in a cause-free reaction of our observation instrument.

Thus objectivity would be disturbed if complete causality did not exist; conversely, the representation of objective physical events is a necessary assumption for the carrying through of the idea of strict causality. That complete causality is an indispensable assumption for the possibility of effecting the representation of objective physical events was already clearly recognized by Kant. But there is no justification for concluding from this insight that the complete validity of the principle of causality throughout nature is guaranteed from the outset independently of experimental experience. All that must be established is this — objectivity also is weakened with the renunciation of complete causality.

This is actually the case in atomic physics; we have seen that the principle of continuity can not be given up without objectivity ceasing. And we have learned that continuity ends in atomic and quantum physics. We can consider the appearance of discontinuities in elementary physical processes as the fundamental proposition of modern quantum physics; the atomistic structure of matter may be in terpreted as one facet of this elementary physical discontinuity which also appears in many other forms.

The above examples have already made it clear that in quantum physics a complete causality, of the type we are accustomed to, no longer exists. But we still want to refer particularly to the difference that exists between the quantum theory use of the probability concept and the Boltzmann evaluation of it (f or t h e kin e tic ex pl an ation of h eat proc es s es). In Boltz ma nn 's considerations statistics was a secondary thing; at that time there was no reason to doubt that the motion of each individual atom could fundamentally be calculated precisely in advance.

The pursuit of such fine processes was voluntarily relinquished and statistical considerations were used as an expression of an incomplete (but sufficient for the result desired) observation of the events. Whereas in quantum theory the primary natural laws themselves take the form of probability expressions in this case the statistical concepts are not an expression of the incompleteness of our insight into events, but rather an expression of an indefiniteness existing in nature itself. Nature herself did not determine individual atomic processes in advance; from case to case she executes unpredictable decisions which show fixed regularities only in the statistical average. But these unpredictable decisions of nature are always connected with elementary quantum physics discontinuities. Indeed it is the independent cases of quantum transitions that are not predetermined. Our earlier example of the interference of light for the screen with two openings alre ady demonstrated that quantum physics must also relinquish the idea of objective events.

With characteristic positivist modesty we limited our problem to the problem of how great the probability is that the light quantum in question will be absorbed in a certain point of the photographic plate. Thus two quantum transitions — and that is typical of the whole modern quantum theory -are placed in a static relation with each oth er. We observe the quantum transformation of the light emission from the point source; then we observe the quantum transformation of the absorption of the light quantum in a certain grain of the plate; and we can theoretically calculate in advance the probability of the occurrence of the second elementary act after the incidence of the first. But we cannot extend the picture of an objective event between both processes in the form of the specification of a continuous path which the light quantum must traverse from the first position to the second.

Formerly one might have been inclined to suspect that the principle of the objectivity of physical events was also inseparably connected with the possibility of the quantitative, mathematical comprehension of natural regularities. But we see that is plainly incorrect; everything we try to explain here in words can be expressed in as mathematically clear a form as Galileo- Newton mechanics. We must replace the precalculation of future events on the basis of complete causality possible in the macrophysical domain by a purely probability prediction. The probabilities of quantum physics processes are themselves determined exactly quantitatively; in the last analysis they are subject to precise mathematical laws of great simplicity and of the most comprehensive validity.

Despite that, the Kantian interpretation, that complete causality (as well as objectivity and continuity) of natural processes is an indispensable provision of every physical thought in general, is still correct in the following sense: quantum physics experiments are also always performed with macrophysical apparatus. We need macrophysical (thus functioning according to strict causality) apparatus to be able to make any correct observations at all and to be able to determine regularities in the atomic world.

Classical physics remains the indispensable support from which an advance into the .world of quanta and atoms becomes possible. This can not be altered by the fact that macrophysical laws can naturally be interpreted as results of quantum physics elementary laws.

The laws which govern the motions of macrophysical bodies must naturally result from the laws to which their individual atoms are subject. There is no difficulty involved in the necessity of interpreting the strict causality of macrophysical events as a result of the purely statistical laws for the elementary processes. For that is the essence of conformity to statistical laws— despite the incalculability of the separate event, an exact, predictable result occurs in the total effect of a large number of individual processes. The fact that we thus interpret the laws of atomic physics as the really true natural laws from which macrophysical laws are derived as results does not permit us to overlook the other fact, that the elementary laws of atomic physics include a tangible content only when they are attached to the frame of macrophysics by concrete application. Herein lies in the final analysis the root of the not only historical, but contemporary significance of the Bohr correspondence principle. For this teaches us to understand the meaning and content of quantum physics laws from their relation to macrophysics.

EXPERIMENTAL EVIDENCE

Experimental evidence has shown us in a most comprehensive way and with a variety exceeding all expectations the atomistic structure of all physical substrata; not only matter, but also light (despite its wave nature which it exhibits "on the other side") has a corpuscular make-up. This evidence drives us to remarkable conclusions. How can one observe and investigate individual atoms? We saw that modern experimental technique permits the very satisfactory performance of experiments immediately involved with individual atoms (and therewith to exactly prove the reality of these atoms, without a doubt). Naturally experimental manipulation of individual atoms remains much more difficult and far different from the investigation and measuring of macrophysical, visibly large bodies which are made up of innumerable atoms.

For macrophysical bodies we have scales and other mechanical, optical or electrical measuring instruments at our disposal; but there is a decisive difference for measurements on atoms. We know that every physical substratum, therefore every physical measuring instrument, is composed of atoms — be they material atoms or electrons or light quanta. This destroys all possibilities of using convenient measuring instruments for this research as is done for macrophysical objects or events.

We have already considered how fundamental for our classical physical ideas and methods of perception is the fact that in macrophysical investigations the back-coupling, the influencing of the object by the observation process, can be made negligibly small through the use of sufficiently fine measuring instruments. But if we consider that all measuring instruments themselves consist of atoms (thus can never be made finer and smaller than single atoms) we see that this method of eliminating the reaction of the measuring instrument is barred when the object to be investigated is itself an individual atom (or a structure containing just a few atoms).

There is no longer any possibility of investigating and observing with instruments that are finer than the object in question. Nor is it possible to control the influencing of the object by the measuring instrument and to eliminate it from the result by a corresponding conversion. Thus it must be considered as part of the bargain that fundamentally measurements on atomic objects are always "falsified" in the sense that according to natural law the object experiences a variable interference from the execution of the observation process.

Similarly when we try to observe and psychologically control our own thought processes we are successful up to a certain degree; but the functioning of the observation itself again influences the observed "object" — our own thought process — and with this sets the limits for the possibility of observation. (One can not, e.g, watch through psychological self-observation how one falls asleep, because just that attention of observing prevents one from falling asleep. Also, *e.g.*, the origin of voluntary decisions is disturbed and changed by internal-controlling self-observation.) Similarities are also found in atomic physics — the observed object is influenced by the observation process itself. With Niels Bohr we can say, the separation between observed object and observing subject begins to vanish here.

It is not at all proper to call this influencing of the object by the process of observation a "falsification". For we are not dealing with a disturbing influence which is in any way limited for the present by the current deficiency of our observation technique; these barriers to the possibility of an ideal observation which does not influence the object itself are limited — through the atomistic structure of physical foundations — by natural law. Therefore, we must ascribe to the atomic objects themselves a certain character of "indeterminateness", of "indefiniteness" of their physical behaviour which makes the construction of an objective picture of atomic physical events impossible.

This need not imply that every possibility of exact measurement on atomic objects is doomed to failure. It is possible throughout to make each physical property of an atom the object of precise measurements. If a certain property of an atom is observed exactly there result from the observation process (due to the reaction of the measuring instrument to the object) strong uncontrollable and undefined changes (strong "uncertainties") with regard to other properties of this atom.

By optionally transferring the interference which is necessarily associated with observation of the atom to different properties of the object it is possible to make accurate observations on the properties of the atom undisturbed by the particular reaction. With this "complementarity", as Bohr named it, the new theory can eliminate those apparently hopeless contradictions we first encountered in the dualism of waves and corpuscles and which we met at every step in quantum physics.

This idea of complementarity must be viewed as the most significant result for philosophy that crystallized out of modern physics. It presents an absolutely new scientific way of thinking which is fundamentally different from classical scientific thinking in terms of objectivized representations. After the intellectual penetration and comprehension of atomic physical phenomena completely inaccessible to the previous method of representation became possible through it, it appears justified to believe that it may further become of epoch-making importance in other realms of natural science.

The enlightening force of this idea in solving an apparently insoluble puzzle and contradiction is clearly demonstrated in the famous problem of the dualistic nature of light. The properties connected with the wave nature of light on the one hand and those connected with its corpuscular nature on the other are "complementary" to each other in the sense that they can never appear in one and the same experiment at the same time (and thus come into actual direct opposition). Experiments which let the wave side of light emerge clearly force (through the action that is connected with every experiment) the corpuscular nature of light back into the indeterminate and unobservable; other experiments, which force the corpuscular side of light into prominence, leave undefined and indiscernible all the properties which usually betray to us the wave nature of light.

With this wonderful device of complementarity nature combines in one and the same physical object properties and regularities that contradict each other so that they could never exist directly at the same time. Let us pursue this in somewhat more detail in the already repeatedly discussed example of the interference of light rays transmitted through two screen openings.

If we want to let the interference of these two light rays take place to the wave nature of light we must, as already indicated, relinquish the desire to determine simultaneously through which of the two openings a definite light quantum w as transmitted. We can also c onsider th is event from the "complementary" opposite side. We can undertake to desire to observe through which opening the light quantum was transmitted.

But the n we m us t c onvers ely relinquis h an y desire to obtain a n interference of the two rays. For with the latter intention the only procedure we can follow is to roughly close up one of the screen openings in order to be certain that a transmitted light quantum could actually only have passed through the other opening. There is no experimental possibility of assuring ourselves in any way through which of the openings the light quantum was trans mitted without simultaneously so a lterin g the conditions of the experiment that the interference of the two rays is hindered.

As another example let us consider the charge cloud of the electron in the hydrogen atom. If we fix the position of a definite point with a microscope we must assume an inaccuracy which is at least of the order of magnitude of the wave lengths of visible light — thus much larger than an atom. But nothing hinders our imagining that we have a microscope that is not dependent on visible light, but on X-rays or even rays of much shorter wave length (gamma rays). With this microscope we could accomplish measurements of position on an electron with such great accuracy that positions inside of the charge cloud of the hydrogen atom would also be distinguished.

Now let us place a hydrogen atom under this gamma-microscope and "examine" the inner structure of this atom so that we determine the position of the electron. In the description of this imaginary experiment we must not forget one thing; namely, that the energy of the "gammalight" necessary to illuminate and render the electron visible is concentrated in individual light quanta which possess very high energy because of the minuteness of their wave lengths. The procedure of the "examination" of the electron in the hydrogen atom is as follows: a single energyrich light quantum meets the electron, and reflected by it approaches us through the microscope, showing us the exact position of the electron. But in this process the electron experiences a terrific effect; it is thoroughly dislodged from its former condition by the interaction with the energy-rich light quantum and in most cases we can expect that the electron is completely torn away from the hydrogen nucleus; the atom, thus, is ionised.

We can now explain in a concrete way the meaning of the "charge cloud" around the atomic nucleus calculated according to wave mechanics. We repeat the a bove des cribed e xperim ent inn umerable times ; and, it m us t be emphasized, each time we use a hydrogen atom which is in its lowest energy state (normal state). If, instead, we took atoms in a fixed, higher energy level we would have to reckon with a different structure of the charge cloud; for each of the different energy levels, according to quantum theory, there exists a certain specific structure of the charge cloud.

By repeatedly measuring the position of the electron in a hydrogen atom in its normal state we find the electron in different positions from case to case, just as in our interference experiment we found the individual light quanta in different places on the photographic plate, statistically distributed according to the light intensity calculated by the wave theory. The statistical distribution of the individually measured electron positions is given by the charge cloud calculated according to wave mechanics. It was only with this determination that the concept of this charge cloud attained a clear meaning which is defined by concrete, demonstrable experiments.

At the same time we see in this experiment how the changing actions in quantum physical observation are bound up with the observed facts. Before the act of observation in question the hydrogen atom possesses a definite energy, but in general in these circumstances a definite position of the electron does not exist — the position of the electron is undefined, or only indefinitely defined in the structure of the statistical charge cloud.

It is only through the process of observing its location in the gamma-ray microscope that we force the electron to assume a definite position. Notice, we do not prescribe in which position it should emerge; but we do force it into some definite position and thereby force it to a new crisis; now the electron assumes a definite position but simultaneously an indefinite energy exchange has taken place between the gamma light quantum and the electron and the original condition of a defined energy of the atom has been disturbed.

A quite analogous process is represented by the previously described impinging of a linearly polarized light quantum against a Nicol prism. In this case we can also say that the execution of an act of observation has forced the light quantum to assume a clearly defined situation, when previously this was indefinite. The light quantum must decide whether to be transmitted by the Nicol prism which has been placed inclined to the original plane of vibration of the light quantum or to be reflected by it. That is a decision of the same kind as that of assuming a definite position forced upon the electron in the above experiment.

It is obvious that there is no more place among these ideas for a complete causality, clearly predetermining each occurrence. If in a macrophysical structure, the planetary system, for example, we want to pre-calculate future movements exactly we must know two things. First, we must know that the Newtonian law is valid (and not any other one), and we must know the magnitudes of the different planets which are determinative for Newtonian gravitational attraction and for the relations of force and acceleration.

Secondly, for any one point in time we must know the positions and velocities the different planets possess at exactly this point in time. Thereby the general course of motion is mathematically precisely determined for all later (besides, also earlier) times. In an electron, however, we are not able to simultaneously determine its position and velocity at a definite point in time. Since position and velocity are complementary, the position measurement in the gamma-microscope makes the velocity of the electron unobservable — and the converse also holds.

After knowing that the physical properties of an atom are partially complementary to one another, that therefore it is impossible to observe the atom simultaneously "from all sides" as it were (as is possible in macrophysical bodies), it must be regarded as quite natural that precalculations of the future conduct of atoms, electrons and light quanta are not possible analogously as in the planets. Let us emphasize once again: this impossibility not only depends on a practical technical incompleteness of our instruments, but depends on nature itself. It is a positive result of the natural laws which in quantum or wave mechanics have attained a formulation which is mathematically exact and is verified by innumerable experiments.

The indefiniteness inherent in the physical condition of all atomic objects stipulates a corresponding indefiniteness in the process of the action; pre- calculation according to exact causal laws is lacking here. The inability of the physicist to predict for an individual case which of the various possibilities will be realized in the quantum transition which is the basis for an observation process is not due to human imperfection of knowledge; nature herself has reserved until the last the decision for each individual case.

Finally, letusmake it clear that our repeated use of the word "indefiniteness" in relation to atomic physical events actually expresses nothing but the impossibility of using familiar classical concepts in the place in question. The impossibility of describing the relations in objectivised process pictures lends as many difficulties to the verbal expression as does the problem of a clear representation.

❒

3.

Matter Waves

INTRODUCTION

We begin our study of quantum mechanics by discussing the diffraction undergone by X-rays and electrons when they interact with a crystal. X-rays are a form of electromagnetic radiation with wavelengths comparable to the distances between atoms. Scattering from atoms in a regular crystalline structure results in an interference pattern which is in many ways similar to the pattern from a diffraction grating. We first develop Bragg's law for diffraction of X-rays from a crystal. Two practical techniques for doing X-ray diffraction are then described.

It turns out that electrons have wave-like properties and also undergo Bragg diffraction by crystals. Bragg diffraction thus provides a crucial bridge between the worlds of waves and particles. With this bridge we introduce the classical ideas of momentum and energy by relating them to the wave vector and frequency of a wave. The properties of waves also give rise to the Heisenberg uncertainty principle.

This gives us a feel for the chronology of these discoveries and indicates how important they were to the development of physics in the early 20th century.

THE WAVE FUNCTION,Y

We've derived a complete many-electron Hamiltonian operator. Of course, the Schrödinger equation involving it is intractable, so let's consider a simpler problem, involving the one-electron hamiltonian

$$\hat{h}(i) = -\frac{1}{2}\nabla_i^2 + \sum_{A}^{M} \frac{1}{r_{iA}}$$

which involves no electron-electron interaction. This soluable in the B-O approximation (recall the hydrogen atom by letting M=1). Call the solutions to the one-electron Schrödinger equation $X i (x)$. These will be molecular spin orbitals when we get around to it, but for now let it suffice to know they satisfy the eigenequation

$$\hat{h}(i)X_j(x_i)=\varepsilon_j X_j(x_i)$$

with the interpretation that electron i occupies spin orbital x_j with energy e_j. If we ignore electron-electron interaction in $\hat{H}_e$, we construct a simpler system with Hamiltonian

$$\hat{H}=\sum_i^N \hat{h}(i)$$

It will have eigenfunctions which are simple products of occupied spin orbitals, and thus an energy which is a sum of individual orbital energies, as

$$\psi^{HP}=X_i(x_1)X_j(x_2)X_k(x_3)...X\eta(x_N)$$

$$\langle\psi|\hat{H}|\psi\rangle=\varepsilon_i+\varepsilon_j+\varepsilon_k+...+\varepsilon\eta=E$$

This kind of wavefunction is called a Hartree Product, and it is not physically realistic. In the first place, it is an independent-electron model, and we know electrons repel each other. Secondly, it does not satisfy the antisymmetry principle due to Pauli which states that the sign of the wavefunction must be inverted under the operation of switchin g the coordinates of any two electrons, or

$$\psi(...X_i...X_j...)=-\psi(...X_j...X_i...)$$

Part of the proof of equation acknowledges this is not so for a Hartree Product. To remedy this, first consider a two-electron system, such as helium. Two equivalent Hartree Product wavefunctions for this system are

$$\psi_1^{HP}=X_i(x_1)X_j(x_2)$$

$$\psi_2^{HP}=X_i(x_2)X_j(x_1)$$

Obviously, neither of these is appropriate. However, using the old "by inspection..." trick, we notice that

$$\psi=\frac{1}{\sqrt{2}}\left[X_i(x_1)X_j(x_2)-X_i(x_2)X_j(x_1)\right]$$

does. The mathematical form of this wavefunction can be generated by a determinant of 's,

$$\psi=2^{-1/2}\begin{vmatrix}X_i(x_1) & X_j(x_1)\\ X_i(x_2) & X_j(x_2)\end{vmatrix}$$

The familiar Pauli exclusion principle follows directly from this example. When we attempt to doubly occupy a spin orbital *Xi* and electron 2 in it, what happens?

$$\psi = 2^{-1/2}\begin{vmatrix} X_i(x_1) & X_j(x_1) \\ X_i(x_2) & X_j(x_2) \end{vmatrix}$$

$$= \frac{1}{\sqrt{2}}\left[X_i(x_1)X_i(x_2) - X_i(x_2)X_i(x_1)\right] = 0$$

Equation can be generalized to give the *N* electron Slater determinant

$$\psi \quad = \quad (N!)^{-1/2}\begin{vmatrix} X_i(x_1) & X_j(x_1) & \cdots & X\eta(x_1) \\ X_i(x_2) & X_j(x_2) & \cdots & X\eta(x_2) \\ \vdots & \vdots & \ddots & \vdots \\ X_i(x_N) & Xj(x_N) & \cdots & X\eta(x_N) \end{vmatrix}$$

A shorthand notation for a Slater determinant has been introduced, where all the diagonal elements in the determinant are written in order as a "ket" vector. Equation can thus be written as

$$\psi = \quad \left|X_i(x_1)X_j(x_2)...X\eta(x_n)\right|$$

where the normalization constant is absorbed into the notation. Now we have written down a wave function appropriate for use in the case where $\hat{H} = \sum \hat{h}(i)$. In HF theory, we make some simplifications so many-electron atoms and molecules be treated this way. By tacitly assuming that each electron moves in a perceived electric field generated by the stationary nuclei and the average spatial distribution of all the other electrons, it essentially becomes an independent-electron problem. The HF Self Consistent Field procedure (SCF) will be bent on constructing each X (x) to give the lowest energy.

WAVEFUNCTION COLLAPSE

Schrödinger's wave equation with its unique wavefunction for a single electron is also spread out in a probability distribution like Heisenberg's quantized particle-like electron. This is because a wave is naturally a widespread disturbance and not a point particle. Therefore, Schrödinger's wave equation has the same predictions made by the uncertainty principle because uncertainty of location is built into the definition of a widespread disturbance like a wave. Uncertainty only needed to be defined from Heisenberg's matrix mechanics because the treatment was from the particle- like aspects of the electron.

Schrödinger's wave equation shows that the electron is in the probability cloud at all times in its probability distribution as a wave that is spread out. Max Born dis covered in 1928 that wh en the square of S ch rödinger's wavefunction (psi-squared) is computed the electron's location as a probability distribution is obtained.

Therefore, to the extent that a measurement of the position of an electron can be made at an exact location instead of as a probability distribution, the electron appears to momentarily cease to have wave-like properties. Without wave-like properties, none of Schrödinger's definitions of the electron being wave-like makes sense. The measurement of the position of the particle nullifies the simple wave-like properties and the one body form of Schrödinger's equation then fails.

Bec ause the elec tron ca n no lon ger be des cribed by its se pa ra te wavefunction when measured, due to its wave length becoming much shorter and its becoming entangled with the particles of the measuring apparatus, this is called wavefunction collapse.

Eigenstates and Eigenvalues

The term eigenstate is derived from the German/Dutch word *"eigen,"* which means "inherent" or "characteristic." The word eigenstate is descriptive of the measured state of some object that possesses quantifiable characteristics such as position, momentum, etc. The state being measured and described must be an "observable" (*i.e.* something that can be experimentally measured either directly or indirectly like position or momentum), and must have a definite value. In the everyday world, it is natural and intuitive to think of everything being in its own eigenstate.

Everything appears to have a definite position, a definite momentum, a definite value of measure, and a definite time of occurrence. However, quantum mechanics affirms that it is impossible to pinpoint exact values for the momentum of a certain particle like an electron in a given location at a particular moment in time, or, alternatively, that it is impossible to give an exact location for such an object when the momentum has been measured. Due to the uncertainty principle, statements regarding both the position and momentum of particles can only be given in terms of a range of probabilities, a "probability distribution".

Eliminating uncertainty in one term maximizes uncertainty in regard to the second parameter. Therefore it became necessary to have a way to clearly formulate the difference between the state of something that is uncertain in the way just described, such as an electron in a probability cloud, and effectively contrast it to the state of something that is not uncertain, something that has a definite value. When something is in the condition of being definitely "pinned-down" in some regard, it is said to possess an eigenstate. For example, if the position of an electron has been made definite, it is said to have an eigenstate of position.

A definite value, such as the position of an electron that has been successfully located, is called the eigenvalue of the eigenstate of position. The German word *"eigen"* was first used in this context by the mathematician David Hilbert in 1904.

Schrödinger's wave equation gives wavefunction solutions, meaning the possibilities where the electron might be, just as does Heisenberg's probability distribution. As stated above, when a wavefunction collapse occurs because something has been done to locate the position of an electron, the electron's state becomes an eigenstate of position, meaning that the position has a known value.

THE PAULI EXCLUSION PRINCIPLE

The Pauli Exclusion Principle states that no electron (or other fermion) can be in the same quantum state as another within an atom. Wolfgang Pauli developed the Exclusion Principle from what he called a "two-valued quantum degree of freedom" to account for the observation of a doublet, meaning a pair of lines differing by a small amount (*e.g.*, something on the order of 0.15Å), in the spectrum of the hydrogen atom. The observation meant that there was more energy in the electron orbital from magnetic moment than had previously been described. In early 1925, the young physicists Uhlenbeck and Goudsmit introduced a theory that the electron rotates in space in the same way that the earth rotates on its axis.

This would account for the missing magnetic moment and allow for two electrons in the same orbital to be different if their spins were in opposite directions, thus satisfying the Exclusion Principle. A new quantum number was then needed, one to represent the momentum involved in the rotation of each electron.

One concise statement of Pauli's principle is as follows:

- There cannot exist an atom in such a quantum state that two electrons within have the same set of quantum numbers.

  An electron has four quantum numbers:

- n, the total quantum number of the orbit
- l, the azimuthal quantum number
- ml, the orbital magnetic quantum number
- ms, the spin magnetic quantum number

"In the case of a helium atom with two electrons in the 1*s* orbital, the Pauli exclusion principle requires that the two electrons differ in the value of one quantum number. Their values of n, l, and ml are the same; moreover, they have the same spin quantum number, s = 1/2. Accordingly they must differ in the value of ms, which can have the value of +½ for one electron and –½ for the other."

Dirac wave Equation

In 1928, Paul Dirac extended the Pauli equation, which described spinning electrons, to account for relativity. By using the simplest electromagnetic interaction, he was able to predict the value of the magnetic moment associated with the electron's spin, and found the experimentally observed value, a value too large to be that of a spinning classical charged sphere.

He was able to solve for the spectral lines of the Hydrogen atom, reproducing Sommerfeld's successful formula for the fine-structure splitting of the spectral lines from sound physical principles. Dirac's theory suffered from a problem of negative energy, which required Dirac to introduce antiparticles and a dynamical vacuum and led to many-particle quantum field theory.

All of the above development of quantum theory was based mainly on the atomic spectrum of the hydrogen atom. This is due to the fact that each atom of each element produces a unique pattern of spectral lines when light from each different kind of element is passed through a prism. Scientists could not study the electron and nucleus of the atom itself because they cannot be seen.

Even to day with High- resolution Scanning Tunnelling Electron Microscopes we can only get images of the atom as a blurry fuzzball. However, the spectral lines of the atom reveal the orbits of electrons and the energy that can be expected. It was basically this study of the spectroscopic analysis of first the hydrogen atom and then the helium atom that led to quantum theory. Therefore, the mathematical formulas were made to fit the picture of the atomic spectrum. That is why quantum mechanics is sometimes referred to as a form of mathematical physics.

Quantum Entanglement

Albert Einstein rejected Heisenberg's Uncertainty Principle insofar as it seemed to imply more than a necessary limitation on human ability to actually know what occurs in the quantum realm. In a letter to Max Born in 1926, Einstein famously declared that "God does not play dice". The bare surface level prescription for making predictions from quantum mechanics, based on Born's rule for computing probabilities, became known as the Copenhagen Interpretation of quantum mechanics. Bohr spent many years developing and refining this interpretation in light of Einstein's objections. After the 1930 Solvay conferen ce, Ein stein never a ga in challen ge d the Copenh agen interpretation on technical grounds, but did not cease a philosophical attack on the interpretation, on the grounds of realism and locality (*i.e.*, the idea that there is no action at a distance).

Einstein, in trying to show that quantum theory was not a complete theory, recognized that the theory predicted that two or more particles that had interacted in the past can exhibit strong correlations later on when various measurements are made on them. He wanted this seeming interaction to be explained in a classical way through their common past, and preferably not by a "spooky action at a distance". The argument is worked out in a famous paper devoted to what is now called the EPR paradox. Assuming what is now usually called "local realism", the EPR attempts to show from quantum theory that particles simultaneously possess both position and momentum, while according to the Copenhagen interpretation, only one of these two properties actually exists, and only in the moment that it is being measured.

Einstein considered this conclusion a proof that quantum theory is incomplete in that it refuses to physical properties which objectively exist in nature. The feature of quantum theory leading to these paradoxes is called quantum entanglement. Quantum entanglement refers to situations wherein the properties of several discrete objects cannot be described simply by considering them separately, even after taking account of the history of their past interaction. A quantum interaction can affect the entangled quantum entity in a like manner even though the two of them are widely separated. Bohr's original response to Einstein was that entangled particles are part of one indivisible system.

Einstein's challenge led to decades of substantial research into quantum entanglement. Research would seem to confirm Bohr's assertion that two or more entangled particles must be viewed together as one whole, and, moreover, that dif ficulties only a rise by in sistin g on the reality of outcom es of measurements that are not made anyway. It might be said that God does indeed throw dice, though rather peculiar ones. A real dice throw can be completely understood using the laws of classical mechanics, and the outcome is merely a function of the initial conditions. However the outcome of tossing quantum "dice" has no antecedent; it has as yet no discernible cause.

INTERPRETATIONS

Despite the success of quantum mechanics, it does have some controversial elements. For example, the behaviour of microscopic objects described in quantum mechanics is very different from our everyday experience, which may provoke some degree of incredulity. Most of classical physics is now recognized to be composed of special cases of quantum physics theory and/or relativity theory. Dirac brought relativity theory to bear on quantum physics so that it could properly deal with events that occur at a substantial fraction of the speed of light. Classical physics, however, also deals with mass attraction (gravity), and no one has yet been able to bring gravity into a unified theory with the relativized quantum theory. Also interesting, according to the correspondence principle and Ehrenfest's theorem as a system becomes larger or more massive (action >> h) the classical dynamics tends to emerge (with some exceptions such as superfluidity).

This is why we can usually ignore quantum mechanics when dealing with everyday objects; instead the classical description will suffice. Even so, trying to make sense of quantum theory is an ongoing process that has spawned a number of interpretations of quantum theory, ranging from the conventional Copenhagen Interpretation to hidden variables and many worlds interpretations. There seems to be no end in sight to the philosophical musings on the subject; however the empirical or technical success of the theory is unrivaled; all modern fundamental physical theories are quantum theories, relativity being subsumed within quantum field theories.

CLASSICAL WAVE EQUATIONS

A fairly brief review of waves in various shaped elastic media—beginning with a taut string, then going on to an elastic sheet, a drumhead, first of rectangular shape then circular, and finally considering elastic waves on a spherical surface, like a balloon. The reason we look at this material here is that these are "real waves", hopefully not too difficult to think about, and yet mathematically they are the solutions of the same wave equation the Schrödinger wave function obeys in various contexts, so should be helpful in visualizing solutions to that equation, in particular for the hydrogen atom. We begin with the stretched string, then go on to the rectangular and circular drumheads. We derive the wave equation from $F = ma$ for a little bit of string or sheet. The equation corresponds exactly to the Schrödinger equation for a free particle with the given boundary conditions.

The most important section here is the one on waves on a sphere. We find the first few standing wave solutions. These waves correspond toSchrodinger 's wave function for a free particle on the surface of a sphere. This is what we need to analyse to understand the hydrogen atom, because using separation of variables we split the electron's motion into radial motion and motion on the surface of a sphere.

The potential *only* affects the *radial* motion, so the motion on the sphere is free particle motion, described by the same waves we find for vibrations of a balloon. (There *is* the generalization to complex non-standing waves, parallel to the one-dimensional extension from sin*kx* and cos*kx* to *eikx* and e^{-ikx}, but this does not affect the structure of the equations.)

WAVES ON A STRING

Let's begin by reminding ourselves of the wave equation for waves on a taut string, stretched between $x = 0$ and $x = L$, tension T newtons, density r kg/meter. Assuming the string's equilibrium position is a straight horizontal line (and, therefore, ignoring gravity), and assuming it oscillates in a vertical plane, we use $f(x,t)$ to denote its shape at instant t, so $f(x,t)$ is the instantaneous upward displacement of the string at position x. We assume the amplitude of oscillation remains small enough that the string tension can be taken constant throughout.

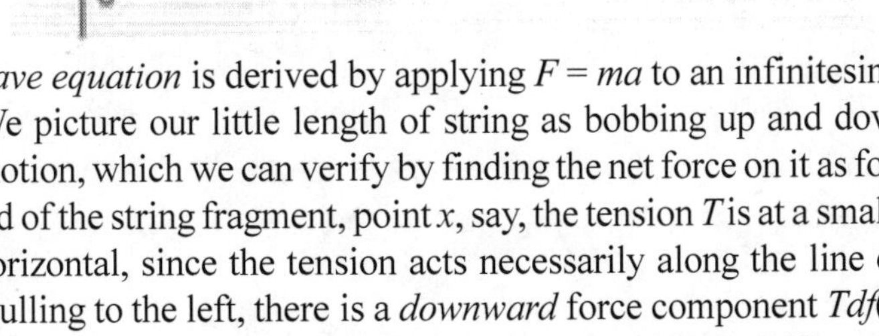

The *wave equation* is derived by applying $F = ma$ to an infinitesimal length dx of string. We picture our little length of string as bobbing up and down in simple harmonic motion, which we can verify by finding the net force on it as follows. At the left hand end of the string fragment, point x, say, the tension T is at a small angle $df(x)/dx$ to the horizontal, since the tension acts necessarily along the line of the string. Since it is pulling to the left, there is a *downward* force component $Tdf(x)/dx$. At the right hand end of the string fragment there is an *upward* force $Tdf(x + dx)/dx$.

Putting $f(x + dx) = f(x) + (df/dx)dx$, and adding the almost canceling upwards and downwards forces together, we find a net force $T(d^2f/dx^2)dx$ on the bit of string. The string mass is $\rho\, dx$, so $F = ma$ becomes

$$T\frac{\partial^2 f(x,t)}{\partial x^2}dx = \rho dx\frac{\partial^2 f(x,t)}{\partial t^2}$$

giving the standard wave equation

$$\frac{\partial^2 f(x,t)}{\partial x^2} = \frac{1}{c^2}\frac{\partial^2 f(x,t)}{\partial t^2}$$

with wave velocity given by $c^2 = T/\text{r}$.

This equation can of course be solved by separation of variables, $f(x,t) = f(x)g(t)$, and the equation for $f(x)$ is identical to the time independent Schrödinger equation for a particle confined to $(0, L)$ by infinitely high walls at the two ends. This is why the eigenfunctions (states of definite energy) for a Schrödinger particle confined to $(0, L)$ are identical to the modes of vibration of a string held between those points. (However, it should be realized that the *time dependence* of the string wave equation and the Schrödinger *time-dependent* equation are quite different, so a non-stationary state, one corresponding to a sum of waves of different energies, will develop differently in the two systems.)

WAVES ON A RECTANGULAR DRUMHEAD

Let us now move up to two dimensions, and consider the analogue to the taut string problem, which is waves in a taut horizontal elastic sheet, like, say, a drumhead. Let us assume a rectangular drumhead to begin with. Then, parallel to the argument above, we would apply $F = ma$ to a small square of elastic with sides parallel to the x and y axes. The tension from the rest of the sheet tugs along all four sides of the little square, and we realise that tension in a sheet of this kind must be defined in newtons per meter, so the force on one side of the little square is given by multiplying this "tension per unit length" by the length of the side.

Following the string analysis, we take the vertical displacement of the sheet at instant t to be given by $f(x, y, t)$. We assume this displacement is quite small, so the tension itself doesn't vary, and that each bit of the sheet oscillates up and down (the sheet is not tugged to one side). Suppose the bottom left- hand corner (so to speak) of the square is (x, y), the top right-hand corner $(x + dx, y + dy)$. Then the left and right edges of the square have lengths dy. Now, what is the total force on the left edge? The force is Tdy, in the local plane of the sheet, perpendicular to the edge dy. Factoring in the slope of the sheet in the direction of the force, the vertically downward component of the force must be $td\partial f(x,y,t)/\partial x$.. By the same argument, the force on the right hand edge has to have an *upward* component Tdy $td\partial f(x,y,t)/\partial x$..

Thus the net upward force on the little square from the sheet tension tugging on its left and right sides is

$$tdy\left(\frac{\partial f(x,dx,y,t)}{\partial x} - \frac{\partial f(x,dx,y,t)}{\partial x}\right) = Tdydx\left(\frac{\partial^2 f}{\partial x^2}\right).$$

The net vertical force from the sheet tension on the other two sides is the same with *x* and *y* interchanged. The mass of the little square of elastic sheet is r*dxdy*, and its upward acceleration is $\partial^2 f / \partial^2$. Thus $F = ma$ becomes:

$$Tdydx\left(\frac{\partial^2 f}{\partial x^2}\right) + Tdydx\left(\frac{\partial^2 f}{\partial y^2}\right) = pdxdy\left(\frac{\partial^2 f}{\partial t^2}\right)$$

giving,

$$\left(\frac{\partial^2 f}{\partial x^2}\right) + \left(\frac{\partial^2 f}{\partial y^2}\right) = \frac{1}{c^2}\left(\frac{\partial^2 f}{\partial t^2}\right)$$

with $c^2 = T/\text{r}$. This equation can be solved by separation of variables, and the time independent part is identical to the Schrödinger time independent equation for a free particle confined to a rectangular box.

WAVES ON A CIRCULAR DRUMHEAD

A similar argument gives the wave equation for a circular drumhead, this time in (r, φ) coordinates (we use φ rather than θ here because of its parallel role in the spherical case, to be discussed shortly).

This time, instead of a tiny square of elastic, we take the small area $rdrd\varphi$ bounded by the circles of radius *r* and *r* + *dr* and lines through the origin at angles φ and j + $d\varphi$. Now, the downward force from the tension *T* in the sheet on the inward curved edge, which has length *rd*j, is $rd\varphi f(r,\varphi,t)/\varphi r$.

On putting this together with the upward force from the other curved edge, it is important to realise that the *r* in *Trd*j varies as well as $Trd\varphi$ on going from *r* to *r* + *dr*, so the sum of the two terms is $Td\varphi\partial/\partial r(r\partial f/\partial r)dr$. To find the vertical elastic forces from the straight sides, we need to find how the sheet slopes in the direction perpendicular to those sides.

The measure of length in that direction is not φ, but $r\varphi$, so the slope is $1/r.\partial f/\partial\varphi$, and the net upward elastic force contribution from those sides (which have length *dr*) is $Tdrd\varphi\partial/\partial\varphi(1/r.\partial f/\partial\varphi)$.

Writing $F = ma$ for this small area of elastic sheet, of mass r*rdrd*j, gives then

$$Td\varphi\frac{\partial}{\partial r}r\frac{\partial f}{\partial r}dr + Tdrd\varphi\frac{\partial}{\partial\varphi}\frac{1}{r}\frac{\partial f}{\partial\varphi} = prdrd\varphi\frac{\partial f}{\partial t^2}$$

which can be written

$$\frac{1}{r}\frac{\partial}{\partial r}r\frac{\partial f}{\partial r} + \frac{1}{r^2}\frac{\partial^2 f}{\partial\varphi^2} = \frac{1}{c^2}\frac{\partial^2 f}{\partial\varphi t}.$$

This is the wave equation in polar coordinates. Separation of variables gives a radial equation called Bessel's equation, the solutions are called Bessel functions. The corresponding electron standing waves have actually been observed for an electron captured in a circular corral on a surface.

WAVES ON A SPHERICAL BALLOON

Finally, let us consider elastic waves on the surface of a sphere, such as an inflated spherical balloon. The natural coordinate system here is spherical polar coordinates, with θ measuring latitude, but counting the north pole as zero, the south pole as π. The angle φ measures longitude from some agreed origin. We take a small elastic element bounded by longitude lines φ and $\varphi + d\varphi$ and latitude θ and $\theta + d\theta$. For a sphere of radius *r*, the sides of the element have lengths $r\sin\theta d\varphi, rd\theta$ etc. Beginning with one of the longitude sides, length $rd\theta$, tension *T*, the only slightly tricky point is figuring its deviation from the local horizontal, which is $1/\sin\theta(\partial f/\partial\varphi)$, since increasing φ by $d\varphi$ means moving an actual distance $r\sin\theta d\varphi$ on the surface, just analogous with the circular case above. Hence, by the usual method, the actual vertical force from tension on the two longitude sides is $Trd\theta d\varphi.\left(\partial/\partial\varphi 1/r\sin\theta.\partial f/\partial\varphi\right)$.

To find the force on the latitude sides, taking the top one first, the slope is given by $1/r.\partial f/\partial\theta$, so the force is just $Tr\text{sinq }\theta d\varphi.1/r.\partial f/\partial\theta$. On putting this together with the opposite side, it is necessary to recall that sinq as well as *f* varies with θ, so the sum is given by: $Trd\varphi d\theta\partial/\partial\theta\sin\theta.1/r,\partial f/\partial\theta$.

We are now ready to write down $F = ma$ once more, the mass of the element is $\rho r^2\sin\theta d\theta d\varphi$. Canceling out elements common to both sides of the equation, we find:

$$\frac{1}{\sin\theta}\frac{\partial}{\partial\theta}\sin\theta\frac{\partial f}{\partial\theta}+\frac{1}{\sin^2\theta}\frac{\partial^2 f}{\partial\varphi^2}=\frac{1}{c^2}\frac{\partial^2 f}{\partial t^2}.$$

Again, this wave equation is solved by separation of variables. The time-independent solutions are called the *Legendre* functions. They are the basis for analysing the vibrations of any object with spherical symmetry, for example a planet struck by an asteroid, or vibrations in the sun generated by large solar flares.

SIMPLE SOLUTIONS TO THE SPHERICAL WAVE EQUATION

Recall that for the two dimensional circular case, after separation of variables the angular dependence was all in the solution to $\partial^2 f/\partial\varphi^2 = -\lambda f$, and the physical solutions must fit smoothly around the circle (no kinks, or it would not satisfy the wave equation at the kink), leading to solutions $\sin m\varphi$ and $\cos m\varphi$ (or $\left(ore^{im\varphi}\right)$)

with m an integer, and $\lambda = m^2$ (this is why we took λ with a minus sign in the first equation).

For the spherical case, the equation containing all the angular dependence is,

$$\frac{1}{\sin\theta}\frac{\partial}{\partial\theta}\sin\theta\frac{\partial f}{\partial\theta}+\frac{1}{\sin^2\theta}\frac{\partial^2 f}{\partial\varphi^2}=\lambda f$$

The standard approach here is, again, separation of variables. Taking the first term on the left hand side over to the right, and multiplying throughout by $\sin^2\theta$ isolates the φ term:

$$\frac{\partial^2 f}{\partial\varphi^2}=-\sin^2\theta\left(\lambda f+\frac{1}{\sin\theta\partial\theta}\sin\theta\frac{\partial f}{\partial\theta}\right)$$

Writing now,

$$f(\theta,\varphi)=f_\theta(\theta)f_\varphi(\varphi)$$

in the above equation, and dividing throughout by f, we find as usual that the left hand side depends only on φ, the right hand side only on θ, so both sides must be constants. Taking the constant as $-m^2$, the φ solution is $e^{\pm im\varphi}$, and one can insert that in the θ equation to give

$$\frac{1}{\sin\theta}\frac{\partial}{\partial\theta}\sin\theta\frac{\partial f_\theta}{\partial\theta}+\frac{m^2}{\sin^2\theta}f_\theta=-\lambda f_\theta$$

What about possible solutions that don't depend on j? The equation would be the simpler,

$$\frac{1}{\sin\theta}\frac{\partial}{\partial\theta}\sin\theta\frac{\partial f}{\partial\theta}=-\lambda f_\theta$$

Obviously, f = constant is a solution (for $m = 0$) with eigenvalue $\lambda = 0$.

Try $f = \cos\theta$. It is easy to check that this is a solution with $\lambda = 2$.

Try $f = \sin\theta$. This is *not* a solution. In fact, we should have realized it cannot be a solution to the wave equation by visualizing the shape of the elastic sheet near the north pole. If $f = \sin\theta$, $f = 0$ at the pole, but rises linearly (for small θ) going away from the pole. Thus the pole is at the bottom of a conical valley. But this conical valley amounts to a kink in the elastic sheet— the slope of the sheet has a discontinuity if one moves along a line passing through the pole, so the shape of the sheet cannot satisfy the wave equation at that point.

This is somewhat obscured by working in spherical coordinated centered there, but locally the north pole is no different from any other point on the sphere, we could just switch to local (x,y) coordinates, and the cone configuration would clearly *not* satisfy the wave equation.

However, $f = \sin\theta\ \sin\theta$ *is* a solution to the equation. It is a worthwhile exercise to see how the j term gets rid of the conical point at the north pole by considering the value of f as the north pole is approached for various values of $\varphi : \varphi = 0, \pi/2, \pi, 3\pi/2$ *say*. The sheet is now smooth at the pole!

We find $f = \sin\theta$ cos, $\sin\theta\ \sin\varphi$ (and so $\sin\varphi\ e^{i\varphi}$) are solutions with $\lambda = 2$. It is straightforward to verify that $f = \cos^2\theta - 1/3$ is a solution with $1 = 6$. Finally, we mention that other $1 = 6$ solutions are $\sin\theta\ \cos\theta \sin\varphi$ and $\sin^2\theta$.

We do not attempt to find the general case here, but we have done enough to see the beginnings of the pattern. We have found the series of eigenvalues 0, 2, 6, ... It turns out that the complete series is given by $\lambda = l(l + 1)$, with $l = 0, 1, 2, ...$ This integer l is the analogue of the integer m in the wave on a circle case. Recall that for the wave on the *circle*, if we chose real wave functions ($\cos m\varphi$, $\sin m\varphi$, not $e^{im\varphi}$) then $2m$ gave the number of nodes the wave had (that is, m complete wavelengths fitted around the circle). It turns out that on the sphere l gives the number of nodal *lines* (or circles) on the surface.

This assumes that we again choose the φ-component of the wave function to be real, so that there will be m nodal circles passing through the two poles corresponding to the zeros of the cos*m*j term. We find that there are $l - m$ nodal *latitude* circles corresponding to zeros of the function of θ.

SUMMARY: FIRST FEW STANDING WAVES ON THE BALLOON

λ	l	m	Form of Solution (unnormalized)
0	0	0	constant
2	1	0	$\cos\theta$
2	1	1	$\sin e^{i\varphi}$
2	1	−1	$\sin\theta e^{-i\varphi}$
6	2	0	$\cos^2\theta - 1/3$
6	2	±1	$\cos\theta\sin\theta\, e^{\pm i\varphi}$
6	2	±2	$\sin^2\theta\, e^{\pm 2i\varphi}$

THE SCHRÖDINGER EQUATION FOR THE HYDROGEN ATOM

In three dimensions, the Schrödinger equation for an electron in a potential can be written:

$$-\frac{\hbar^2}{2m}\left(\frac{\partial^2\psi}{\partial x^2}+\frac{\partial^2\psi}{\partial y^2}+\frac{\partial^2\psi}{\partial z^2}\right)+v(x,y,z)\psi = E\psi$$

This is the obvious generalization of our previous two-dimensional discussion, and we will later be using the equation in the above form to discuss electron wave functions in metals, where the standard approach is to work with standing waves in a rectangular box.

Recall that in our original "derivation" of the Schrödinger equation, by analogy with the Maxwell wave equation for light waves, we argued that the differential wave operators arose from the energy-momentum relationship for the particle, that is,

$$\frac{P^2x + P_y^2 + P_z^2}{2m}\psi = -\frac{\hbar^2}{2m}\left(\frac{\partial^2\psi}{\partial x^2} + \frac{\partial^2\psi}{\partial y^2} + \frac{\partial^2\psi}{\partial z^2}\right) = \frac{\hbar^2\nabla^2\psi}{2m}$$

so that the time-independent Schrödinger wave equation is nothing but the statement that $E = K.E. + P.E.$ with the kinetic energy expressed as the equivalent operator.

To make further progress in solving the equation, the only trick we know is separation of variables. Unfortunately, this won't work with the equation as given above in (x, y, z) coordinates, because the potential energy term is a function of x, y and z in a non-separable form.

The solution is, however, fairly obvious: the potential is a function of radial distance from the origin, independent of direction. Therefore, we need to take as our coordinates the radial distance r and two parameters fixing direction, θ and φ. We should then be able to separate the variables, because the potential only affects radial motion. No potential term will appear in the equations for $\theta\varphi$ motion, that will be free particle motion on the surface of a sphere.

MOMENTUM AND ANGULAR MOMENTUM WITH SPHERICAL COORDINATES

It is worth thinking about what are the natural momentum components for describing motion in spherical polar coordinates (r, θ, φ). The radial component of momentum, p_r, points along the radius, of course. The θ- component p_θ points along a line of longitude, away from the north pole if positive (remember θ itself measures *latitude*, counting the north pole as zero). The φ-momentum component, p_φ, points along a line of latitude.

It will be important in understanding the hydrogen atom to connect these momentum components $(p_r, P_\theta, P_\varphi)$ with the *angular* momentum components of the atom. Evidently, momentum in the r-direction, which passes directly through the centre of the atom, contributes nothing to the angular momentum.

Consider now a particle for which $P_r = P_\theta = 0$, only P_φ being non-zero. Classically, such a particle is circling the north pole at constant latitude θ, say, so it is moving in space in a circle or radius $r\sin\theta$ in a plane perpendicular to the north-south axis of the sphere. Therefore, it has an angular momentum about that axis.

$$P_\varphi r\sin\theta = L_z, say.$$

(The standard transformation from (x, y, z) coordinates to (r,θ,φ) coordinates is to take the north pole of the θ,φ sphere to be on the z-axis.)

The wave equation describing the φ motion is a simple one, with solutions of the form $e^{im\varphi}$ with integer m, just as in the two-dimensional circular well. This just means that *the component of angular momentum along the z-axis is quantized,* $L_z = m\hbar$, *with m an integer*.

TOTAL ANGULAR MOMENTUM AND WAVES ON A BALLOON

The *total* angular momentum is $L = rp\perp$, where $p\perp$ is the component of the particle's momentum perpendicular to the radius, so

$$P_\perp^2 = P_\varphi^2 + P_\theta^2.$$

Thus the square of the total angular momentum is (apart from a constant factor) the kinetic energy of a particle moving freely on the surface of a sphere. The equivalent Schrödinger equation for such a particle is the wave equation given in the last section for waves on a balloon. (This can be established by the standard change of variables routine on the differential operators). Therefore, the solutions we found for elastic waves on a sphere actually describe the angular momentum wave function of the hydrogen atom. We conclude that the total angular momentum is quantized, $L^2 = l(l+1)\hbar^2$, with *l an integer.*

ANGULAR MOMENTUM AND THE UNCERTAINLY PRINCIPLE

The conclusions of our above waves on a sphere analysis of the angular momentum of a quantum mechanical particle are a little strange. We found that the component of angular momentum in the z-direction must be a whole number of ‘ units, yet the square of the total angular momentum $L^2 = l(l+1)\hbar^2$, is not a perfect square! One might wonder if the component of angular momentum in the x-direction isn't also a whole number of ‘ units as well, and if not, why not?

The key is that in questions of this type we are forgetting the essentially wavelike nature of the particle's motion, or, equivalently, the uncertainty principle. Recall first that the z-component of angular momentum, that is, the angular momentum about the z-axis, is the product of the particle's momentum in the xy-plane and the distance of the line of that motion from the origin. There is no contradiction in specifying that momentum and that position simultaneously, because they are in perpendicular directions.

However, we cannot at the same time specify either of the other components of the angular momentum, because that would involve measuring some component of momentum in a direction in which we have just specified a position measurement. We *can* measure the total angular momentum, that involves additionally only the component P_θ of momentum perpendicular to the P_φ needed for the z-component.

Thus the uncertainty principle limits us to measuring at the same time only the total angular momentum and the component in one direction. Note also that if we knew the *z*-component of angular momentum to be *m*h, and the total angular momentum were $L^2 = l^2\hbar^2$ with $l = m$, then we would also know that the *x* and *y* components of the angular momentum were exactly zero.

Thus we would know all three components, in contradiction to our uncertainly principle arguments. This is the essential reason why the square of the total angular momentum is greater th an th e maximum s quare of an y one component. It is as if there were a "zero point motion" fuzzing out the direction.

Another point related to the uncertainty principle concerns measuring just where in its circular (say) orbit the electron is at any given moment. How well can that be pinned down? There is an obvious resemblance here to measuring the position and momentum of a particle at the same time, where we know the fuzziness of the two measurements is related by $\Delta p \Delta x \sim h$. Naïvely, for a circular orbit of radius *r* in the *xy*-plane, $pr = L_z$ and distance measured around the circle is $r\theta$, *so* $\Delta p \Delta x \sim h$ suggests $\Delta L_z \Delta\theta \sim h$.. That is to say, precise knowledge of L_z implies no knowledge of where on the circle the particle is. This is not surprising, because we have found that for $L_z = m\hbar$ the wave has the form $e^{im\varphi}$, and so $|\psi|^2$, the relative probability of finding the particle, is the same anywhere in the circle.

On the other hand, if we have a time-dependent wave function describing a particle orbiting the nucleus, so that the probability of finding the particle at a particular place varies with time, the particle cannot be in a definite angular momentum state. This is just the same as saying that a particle described by a wave packet cannot have a definite momentum.

THE SCHRÖDINGER EQUATION IN (r, θ, φ) COORDINATES

It is worth writing first the energy equation for a classical particle in the Coulomb potential:

$$\frac{1}{2m}\left(P_r^2 + P_\theta^2 + P_\varphi^2\right) - \frac{1}{4\pi\varepsilon_0}\frac{e^2}{r} = E$$

This makes it possible to see, term by term, what the various parts of the Schrödinger equation signify. In spherical polar coordinates, Schrödinger 's equation is:

$$-\frac{\hbar^2}{2m}\left(\frac{1}{r}\frac{\partial^2}{\partial r^2}(r\psi)\frac{1}{r^2}\left\{\frac{1}{\sin\theta}\frac{\partial}{\partial\theta}\left(\sin\theta\frac{\partial\psi}{\partial\theta}\right) + \frac{1}{\sin^2\theta}\frac{\partial^2\psi}{\partial\varphi^2}\right\}\right) - \frac{1}{4\pi\varepsilon_0}\frac{e^2}{r}\psi = E\psi$$

SEPARATING THE VARIABLES: THE MESSY DETAILS

We look for separable solutions of the form

$$\psi(r,\theta,\varphi) = R(r)\Theta(\theta)\Phi(\varphi)$$

We now follow the standard technique. That is to say, we substitute $R\Theta\Phi$ for y in each term in the above equation. We then observe that the differential operators only actually operate on one of the factors in any given part of the expression, so we put the other two factors to the left of these operators. We then divide the entire equation by $R\Theta\Phi$, to get

SEPARATING OUT AND SOLVING THE $\Phi(\varphi)$ EQUATION

The above equation can be rearranged to give:

$$\left(\frac{1}{R}\frac{1}{r}\frac{\partial^2}{\partial r^2}(rR)\left\{\frac{1}{\Theta}\frac{1}{\sin\theta}\frac{\partial}{\partial\theta}\left(\sin\theta\frac{\partial\Theta}{\partial\theta}\right)+\frac{1}{\sin^2\theta}\frac{1}{\Phi}\frac{\partial^2\Phi}{\partial\varphi^2}\right\}\right)$$

$$=\frac{2m}{\hbar^2}\left(E=\frac{1}{4\pi\varepsilon_0}\frac{e^2}{r}\right)$$

Further rearrangement leads to:

$$\frac{1}{\Phi}\frac{\partial^2\Phi}{\partial\varphi^2}=\sin^2\theta\left[\left(r^2\left\{\frac{2m}{\hbar^2}\left(E+\frac{1}{4\pi\varepsilon_o}\frac{e^2}{r}\right)+\frac{1}{R}\frac{1}{r}\frac{\partial^2}{\partial r^2}(rR)\right\}\right)\right.$$

$$\left.-\frac{1}{\Theta}\frac{1}{\sin\theta}\frac{\partial}{\partial\theta}\left(\sin\theta\frac{\partial\Theta}{\partial\theta}\right)\right]$$

At this point, we have achieved the separation of variables! The left hand side of this equation is a function *only of* φ, the right hand side is a function *only of r and* θ. The only way this can make sense is if *both sides of the equation are in fact constant* (and of course equal to each other). Taking the left hand side to be equal to a constant we denote for later convenience by m^2,

$$\frac{\partial^2\Phi(\varphi)}{\partial\varphi^2}=-m^2\Phi(\varphi)$$

We write the constant $-m^2$ because we know that as a factor in a wave function $\Phi(\varphi)$ must be single valued as φ increases through 2π, so an integer number of oscillations must fit around the circle, meaning Φ is sin*m*j, cos*m* φ or $e^{im\varphi}$ with *m* an integer. These are the solutions of the above equation. Of course, this is very similar to the particle in the circle in two dimensions, *m* signifies units of angular momentum about the *z*-axis.

SEPARATING OUT THE Q(Q) EQUATION

Backing up now to the equation in the form

$$\left(\frac{1}{R}\frac{1}{r}\frac{\partial^2}{\partial r^2}(rR)\frac{1}{r^2}\left\{\frac{1}{\Theta}\frac{1}{\sin\theta}\frac{\partial}{\partial\theta}\left(\sin\theta\frac{\partial\Theta}{\partial\theta}\right)+\frac{1}{\sin^2\theta}\frac{1}{\Phi}\frac{\partial^2\Phi}{\partial\varphi^2}\right\}\right)$$

$$=\frac{2m}{\hbar^2}\left(E+\frac{1}{4\pi\varepsilon_o}\frac{e^2}{r}\right)$$

we can replace the

$$\frac{1}{\Phi}\frac{\partial^2\Phi}{\partial\Phi^2}$$

term by $-m^2$, and move the r term over to the right, to give,

$$\frac{1}{\Phi}\frac{1}{\sin\theta}\frac{\partial}{\partial\theta}\left(\sin\theta\frac{\partial\Phi}{\partial\theta}\right)-\frac{m^2}{\sin^2\theta}$$

$$=r^2\left\{\frac{2m}{\hbar^2}\left(E+\frac{1}{4\pi\varepsilon_o}\frac{e^2}{r}\right)-\frac{1}{R}\frac{1}{r}\frac{\partial^2}{\partial r^2}(rR)\right\}$$

We have again managed to *separate the variables* the left hand side is a function *only* of θ, the right hand side a function of r. Therefore both must be equal to the same constant, which we set equal to $-\lambda$.

This gives the $\Phi(\theta)$ equation:

$$\frac{1}{\sin\theta}\frac{\partial}{\partial\theta}\sin\frac{\partial\Theta(\theta)}{\partial\theta}-\frac{m^2}{\sin^2\theta}\Theta(\theta)=-\lambda\Theta(\theta)$$

This is exactly the wave equation we discussed above for the elastic sphere, and the allowed eigenvalues λ are $l(l+1)$, where $l = 0, 1, 2,..$ with $l \geq m$.

THE R(R) EQUATION

Replacing the θ,φ operator with the value found just above in the original Schrödinger equation gives the equation for the radial wave function:

$$\frac{\hbar^2}{2m}\left(\frac{1}{r}\frac{\partial^2}{\partial r^2}(rR(r))-\frac{1(l+1)}{r^2}R(r)\right)-\frac{1}{4\pi\varepsilon_o}\frac{e^2}{r}R(r)=ER(r)$$

The first term in this radial equation is the usual radial kinetic energy term, equivalent to $P_r^2 / 2m$ in the classical picture. The third term is the Coulomb potential energy. These condterm is aneffective potential representing the centrifugal force. This is clarified by reconsidering the energy equation for the classical case, 2

$$\frac{1}{2m}\left(P_r^2 + p_\theta^2 + p_\varphi^2\right) - \frac{1}{4\pi\varepsilon_o}\frac{e^2}{r} = E$$

The Angular momentum squard,

$$L^2 = r^2\left(p_\theta^2 + p_\varphi^2\right) = l(l+1)\hbar^2$$

Thus for fixed angular momentum, we can write the above "classical" equation as

$$\frac{1}{2m}\left(p_r^2 + \frac{l(l+1)\hbar^2}{r^2}\right) - \frac{1}{4\pi\varepsilon_o}\frac{e^2}{r} = E.$$

The parallel to the radial Schrödinger equation is then clear.

We must find the solutions of the radial Schrödinger equation that decay for large r. These will be the bound states of the hydrogen atom. In natural units, measuring lengths in units of the first Bohr radius, and energies in Rydberg units

$$r = a_o\rho = \frac{4\pi\varepsilon_o\hbar^2}{me^2}\rho,\ E = E_R\varepsilon\frac{me^4}{2\hbar^2\left(4\pi\varepsilon_o\right)^2}\varepsilon.$$

Finally, taking $u(r) = rR(r)$, the radial equation becomes,

$$-\frac{d^2u(\rho)}{d\rho^2} + \frac{l(l+1)}{\rho^2}u(\rho) - \frac{2}{\rho}u(\rho) = \varepsilon u(\rho).$$

WAVE AND PARTICLE QUANTITIES

Let us now recapitulate what we know about relativistic waves, and how this knowledge translates into knowledge about the mass, energy, and momentum of particles. In the following equations, the left form is expressed in wave terms, *i.e.*, in terms of frequency, wavenumber, and rest frequency. The right form is the identical equation expressed in terms of energy, momentum, and mass. Since the latter variables are just scaled forms of the former, the two forms of each equation are equivalent.

We begin with the dispersion relation for relativistic waves:

$$w^2 = k^2c^2 + \mu^2$$

$$E^2 = \Pi^2c^2 + m^2c^4$$

Calculation of the group velocity, $ug = dw / dk$, from the dispersion relation yields,

$$u_g = \frac{c^{2k}}{w}$$

$$u_g = \frac{c^2\Pi}{E}$$

These two sets of equations represent what we know about relativistic waves, and what this knowledge tells us about the relationships between the mass, energy, and momentum of relativistic particles. When in doubt, refer back to these equations, as they work in all cases, including for particles with zero mass!

It is useful to turn equations around so as to express the frequency as a function of rest frequency and group velocity,

$$w = \frac{\mu}{\left(1 - u_g^2 / c^2\right)^{1/2}}$$

$$E = \frac{mc^2}{\left(1 - u_g^2 / c^2\right)^{1/2}}$$

and the wavenumber as a similar function of these quantities:

$$K = \frac{\mu u_g / c^2}{\left(1 - u_g^2 / c^2\right)^{1/2}}$$

$$\Pi = \frac{m u_g}{\left(1 - u_g^2 / c^2\right)^{1/2}}$$

Note that equations work only for particles with non-zero mass! For zero mass particles you need to use equations with $m = 0 \textit{ and } \mu = 0$

The quantity w - μ indicates how much the frequency exceeds the rest frequency. Notice that if w = m , then from equation $k = 0$. Thus, positive values of $k = w$ - μ indicate $k > 0$, which means that the particle is moving according to equation. Let us call k the kinetic frequency:

$$k = \left[\frac{1}{\left(1 - u_g^2 / c^2\right)^{1/2}} - 1\right]\mu$$

$$k = \left[\frac{1}{\left(1 - u_g^2 / c^2\right)^{1/2}} - 1\right]mc^2$$

We call K the kinetic energy for similar reasons. Again, equation only works for particles with non-zero mass. For zero mass particles the kinetic energy equals the total energy.

Non-Relativistic Limits

When the mass is non-zero and the group velocity is much less than the speed of light, it is useful to compute approximate forms of the above equations valid in this limit.

Using the approximation $(1+\in)^x \approx 1 + x \in$, we find that the dispersion relation becomes,

$$w = \mu + \frac{k^2c^2}{2\mu}$$

$$E = mc^2 + \frac{\Pi^2}{2m}$$

and the group velocity equation takes the approximate form,

$$u_g = \frac{c^2k}{\mu}$$

$$u_g = \frac{\Pi}{m}$$

The non-relativistic limits for equations become,

$$w = \mu + \frac{\mu u_g^2}{2c^2}$$

$$E = mc^2 + \frac{mu_g^2}{2}$$

and,

$$k = \mu u_g / c^2$$

$$\Pi = mu_g$$

while the approximate kinetic energy equation is,

$$k = \frac{\mu u_g^2}{2c^2}$$

$$K = \frac{\mu u_g^2}{2}$$

Just a reminder — the equations in this section are not valid for massless particles.

PLANE WAVE SOLUTIONS

The best way to gain understanding of Schrodinger's equation is to solve it for various potentials. The simplest is a one-dimensional "particle in a box" problem. The appropriate potential is $V(x) = 0$ for x between 0, L and $V(x)$ = infinity otherwise—that is to say, there are infinitely high walls at $x = 0$ and $x = L$, and the particle is trapped between them. This turns out to be quite a good approximation for electrons in a long molecule, and the three-dimensional version is a reasonable picture for electrons in metals. Between $x = 0$ and $x = L$ we have $V = 0$, so the wave equation is just

$$i\hbar\frac{\partial\psi(x,t)}{\partial t} = -\frac{\hbar^2\partial^2\psi(x,t)}{2m \quad \partial x^2}$$

A possible plane wave solution is,

$$\psi(x,t) = Ae^{\frac{i}{\hbar}(px-bt)}.$$

On inserting this into the zero-potential Schrödinger equation above we find $E = p^2/2m$, as we expect. It is very important to notice that the complex conjugate, proportional to,

$$-\frac{i}{e^{\hbar}}(px - Bt)$$

is *not* a solution to the Schrödinger equation! If we blindly put it into the equation we get, an unphysical result.

$$E = -p^2/2m,$$

However, a wave function proportional to,

$$-\frac{i}{e^{\hbar}}(px - Bt)$$

and,

$$-\frac{i}{e^{\hbar}}(px - Bt)$$

respectively.

Note that these two solutions have the *same time dependence*

$$-\frac{iBt}{e^{\hbar}}.$$

To decide on the appropriate solution for our problem of an electron in a box, of course we have to bring in the walls—what they mean is that $\psi = 0$ for $x < 0$ and for $x > L$ because remember $|\psi|^2$ tells us the probability of finding the particle anywhere, and, since it's in the box, it's trapped *between* the walls, so there's zero probability of finding it outside. The condition y = 0 at $x = 0$ and $x = L$ reminds us of the vibrating string with two fixed ends—the solution of the string wave equation is standing waves of sine form. In fact, taking the difference of the two permitted plane-wave forms above gives a solution of this type:

$$y(x,t) = A\sin\frac{px}{\hbar}e^{\frac{iBt}{\hbar}}$$

Thus the allowed values of p are $hn/2L$, where $n = 1, 2, 3\ldots$, and from $E = p^2/2m$ the allowed energy levels of the particle are:

$$E = \frac{p^2}{2m} = \frac{1}{2m}\left(\frac{h}{2L}\right)^2, \frac{4}{2m}\left(\frac{h}{2L}\right)^2, \frac{9}{2m}\left(\frac{h}{2L}\right)^2, \ldots$$

Note that these energy levels become more and more widely spaced out at high energies, in contrast to the hydrogen atom potential. (The harmonic oscillator potential gives equally spaced energy levels, so by studying how the spacing of energy levels varies with energy, we can learn something about the shape of the potential.)

What about the overall multiplicative constant A in the wave function? This can be real or complex. To find its value, note that at a fixed time, say $t = 0$, the probabili ty of the el ectron being betw een x and $x + dx$ is $|\psi|^2\, dx$ or

$$|A|^2 \sin^2 \frac{px}{\hbar} dx.$$

The *total* probability of the particle being *somewhere* between 0, L must be unity:

$$\int_{x-0}^{x-L} |A|^2 \sin^2 \frac{px}{\hbar} dx = 1, so \frac{1}{2} L |A|^2 = 1.$$

Hence,

$$\psi(x,t) = \sqrt{\frac{2}{L}} \sin \frac{px}{\hbar} e^{-\frac{iBt}{\hbar}}.$$

When A is fixed in this way, by demanding that the total probability of finding the particle somewhere be unity, it is called the normalization constant.

STATIONARY STATES

Notice that at a later time the probability distribution for the wave function

$$\psi(x,t) = \sqrt{\frac{2}{L}} \sin \frac{px}{\hbar} e^{-\frac{iBt}{\hbar}}.$$

is the same, because time only appears as a phase factor in this time-dependent function, and so does not affect $|\psi|^2$.

A state with a time-independent probability distribution is called a stationary state.

States with Moving Probability Distributions

Recall that the Schrödinger equation is a linear equation, and the sum of any two solutions is also a solution to the equation. That means that we can add two solutions having different energies, and still have a legal wave function. We shall establish that in this case, the probability distribution varies in time. The simplest way to see how this must be is to look at an example. Let's add the ground state to the first excited state, and normalize the sum:

$$\psi(x,t)=\sqrt{\frac{1}{L}}\left(\sin\frac{\pi x}{L}e^{-\frac{izkt}{4mL^2}}+\sin\frac{2\pi x}{L}e^{-\frac{izkt}{mL^2}}\right).$$

note: h, not $\hbar$. (You can check the normalization constant at $t = 0$). For general x, the two terms in the bracket rotate in the complex plane at different rates, so their sum has a time-varying magnitude.

That is to say, $\left|\psi(x,t)^2\right|$ varies in time, so the particle must be moving around this is *not* a stationary state.

***Exercise*:** To see this, note that at $t = 0$ the wave function is:

$$\psi(x,0)=\sqrt{\frac{1}{L}\left(\sin\frac{\pi x}{L}+\sin\frac{2\pi}{L}\right)}$$

and sketch this function: the particle is more likely to be found in the left-hand half of the box.

Now, suppose the time is $t = 4mL^2 / h, so e^{-\frac{izkt}{4mL-2}} = 1$. At this time

$$\psi\left(4mL^2/h\right)=\sqrt{\frac{1}{L}}\left(-\sin\frac{\pi x}{L}+\sin\frac{2\pi x}{L}\right)$$

and it's easy to see that the particle is more likely to be found n the *right*-hand half.

That is to say, this wave function, a linear sum of wave functions corresponding to different energies, has a probability distribution that sloshes back and forth in the box: and, any attempt to describe a classical-type particle motion, bouncing back and forth, *necessarily* involves adding quantum wave functions of different energies.

Note that the frequency of the sloshing motion depends on the *difference* of the two energies: how constructively the two components interfere depends on the difference of the phases in the energies at the time. A single energy wave function always has a static probability distribution. Of course, the *total* probability of finding the particle *somewhere* in the box remains unity: the normalization constant is time-independent.

THE WAVE-PARTICLE PUZZLE

De Broglie's explanation of the Bohr atom quantization rules, together with the accidental discovery of electron diffraction scattering by Davisson and Germer, make a very convincing case for the wave nature of the electron. Yet the electron different from the velocity of the electron. The point is that the electron waves, like water waves but unlike electromagnetic waves, have differing phase and group velocities. To see this, consider again the beating of two waves of slightly different wavelengths.

$$\sin\left(\left(k-\Delta k\right)x-\left(w-\Delta\omega\right)t\right)\sin\left(\left(k+\Delta k\right)x-\left(\omega+\Delta\omega\right)t\right)$$
$$=2\sin\left(kx-\omega t\right)\cos\left(\left(\Delta k\right)x-\left(\Delta\omega\right)t\right)$$

The waves described by the term sin $\left(kx-\omega t\right)$ have velocity ½v, as previously derived. But the envelope, the shape of the wave packet, has velocity $\Delta\omega / \Delta k$ rather than ω / k. These velocities would be the same if w were linear in ω, as it is for ordinary electromagnetic waves. But the $\omega - k$ relationship follows from the energy-momentum relationship for the (non-relativistic) electron, $E = ½mv^2 = p^2/2m$.

So $so\ dE / dp = p / m = v$. But $E = hf = h\omega / 2p, and\ p = h / \lambda = hk / 2\pi$. Therefore, $\Delta\omega / \Delta k = dE / dp = v$. So the packet travels at the speed we know the electron must travel at, even though the wave peaks within the packet travel at one-half the speed.

SUPERPOSITION OF MECHANICAL WAVES

Two or more waves can pass simultaneously through the same region of space. We can ask what the result of this is. Waves follow the principle of superposition which says that the net effect of the waves is the sum of their individual effects. Overlapping waves do not alter the travel of each other. The mathematics for describing overlapping waves of the same frequency and wavelength, we call this interference, is not difficult. For example, imagine tw o tra ns verse w av es with the s am e amplitude, freq uen cy, direction, and wavelength traveling along a string. We can specify different spatial positions for these waves through different phase angles.

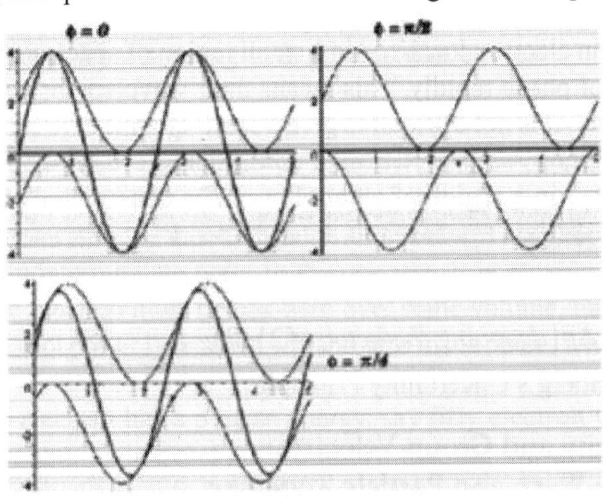

Fig. Two Waves with the Same Amplitude and Frequency but Independent Phase Angles are Depicted in Blue.

If we plot this equation, we see that it is also a sinusoidal wave traveling in the +x direction. This net wave is the *only* wave you would see traveling along the string. The two interfering waves would not be visible. We note that by changing the value of f we can get a resultant wave which shows constructive interference or destructive interference. For completely constructive interference, we want crests of one wave to coincide with crests of the other and troughs to coincide with troughs. For completely destructive interference, we want crests of one wave to coincide with troughs of the other, thereby leading to complete cancellation of the net wave. We see examples of these cases in figure.

The top curve has phase angle equal to zero and the bottom curve has the phase angle set as shown by the label of the graph.The red curve is the result of superposing the two waves. We now come to the first hard part of the class. We'll try to learn what insights we can gain from the equation governing relativistic quantum mechanics.

The Difference Between Relativistic and NR QM

One of the key points in particles physics is that special relativity plays a key rôle. As you all know, in ordinary quantum mechanics we ignore relativity. Of course people attempted to generate equations for relativistic theories soon after Schrodinger wrote down his equation. There are two such equations, one called the Klein-Gordon and the other one called the Dirac equation. The structure of the ordinary Schrodinger equation of a free particle (no potential) suggests what to do.

We can write this equation as,

$$\hat{H}\psi = \frac{1}{2m}p^2\psi = t\hbar\frac{\partial}{\partial t}\psi.$$

This is clearly a statement of the non-relativistic energy-momentum relation,

$$E = \frac{1}{2}mv^2,$$

since a time derivative on a plane wave brings down a factor energy. Remember, however, that p as an operator also contains derivatives,

$$E = \frac{\hbar}{i}\nabla.$$

A natural extension would to use the relativistic energy expression,

$$\hat{H}\psi = \sqrt{m^2c^4 + p^2c^2\psi} = i\hbar\frac{\partial}{\partial t}\psi$$

But this is a non-sensical equation, unless we specify how to take the square root of the operator. The first attempt to circumvent this problem, by Klein and Gordon, was to take the square of the equation,

$$\left(m^2c^4 + p^2c^2\right)\psi = -\hbar\frac{\partial^2}{\partial t^2}\psi.$$

This is an excellent equation for spin-less particles or spin one particles (bosons), but not to describe fermions (half-integer spin), since there is no information about spin is in this equation. This needs careful consideration, since spin must be an intrinsic part of a realtivistic equation!

Dirac realised that there was a way to define the square root of the operator. The trick he used was to define four matrices a, b that each have the property that their square is one, and that they anticommute,

$$\alpha i\alpha i = I, \beta\beta = I,$$

$$\alpha i\,\beta + \beta\alpha i = 0,\ \alpha i\ \alpha j + \alpha j\ \alpha i = 0\ i \neq j.$$

This then leads to an equation that is linear in the momenta-and very well behaved,

$$\left(\beta mc^2 + c\ \infty p\right)\psi = i\hbar\frac{\partial}{\partial t}\psi$$

Note that the minimum dimension for the matrices in which we can satisfy all conditions is 4, and thus ψ is a four-vector! This is closely related to the fact that these particles have spin.

Let us investigate this equation a bit further. One of the possible forms of αi, and β is,

$$\alpha i = \begin{pmatrix} 0 & \sigma \\ \sigma l & 0 \end{pmatrix}, \beta = \begin{pmatrix} I & 0 \\ 0 & -I \end{pmatrix},$$

where s are the two-by-two Pauli spin matrices,

$$\sigma_1 = \begin{pmatrix} 0 & 1 \\ 1 & 0 \end{pmatrix}, \sigma_2 = \begin{pmatrix} 0 & -i \\ i & 0 \end{pmatrix}, \sigma_3 = \begin{pmatrix} 1 & 0 \\ 0 & -1 \end{pmatrix}$$

These matrices satisfy some very interesting relations. For instance,

$$\sigma_1\sigma_2 = i\sigma_3, \sigma_2\sigma_1 = -i\sigma_3, \sigma_2\sigma_3 == i\sigma_1,$$

etc. Furthermore s 2 = 1. Once we know the matrices, we can try to study a plane-wave solution,

y(x, *t*) = *u*(p)ei(p. x – Et)/ h .

Just as photons are the best way to analyse (decompose) the electro- magnetic field, electrons and positrons are the natural way way to decompose the Dirac field that is the general solution of the Dirac equation.

This analysis of a solution in terms of the particles it contains is called (incorrectly, for historical reasons) "second quantization", and just means that there is a natural basis in which we can say there is a state at energy E, which is either full or empty. This could more correctly be referred to as the "occupation number representation" which should be familiar from condensed matter physics. This helps us to see how a particle can be described by these wave equations. There is a remaining problem, however!

ANTIPARTICLES

Both the Klein-Gordon and the Dirac equation have a really nasty property. Since the relativistic energy relation is quadratic, both equations have, for every positive energy solution, a negative energy solution.

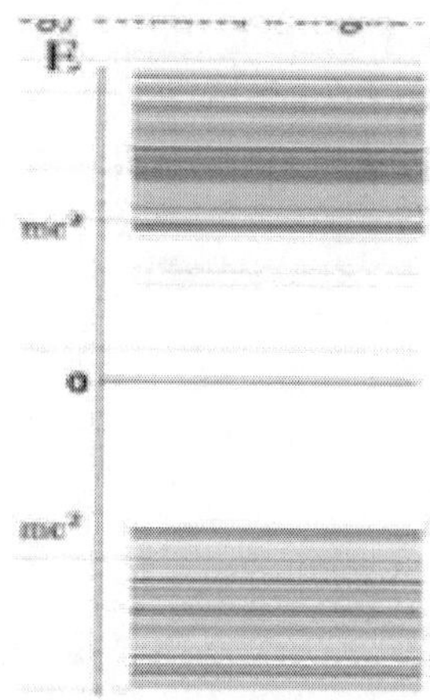

Fig. A Schematic Picture of the Levels in the Dirac Equation.

We don't really wish to see such things, do we? Energies are always positive and this is a real problem. The resolution is surprisingly simple, but also very profound-It requires us to look at the problem in a very different light.

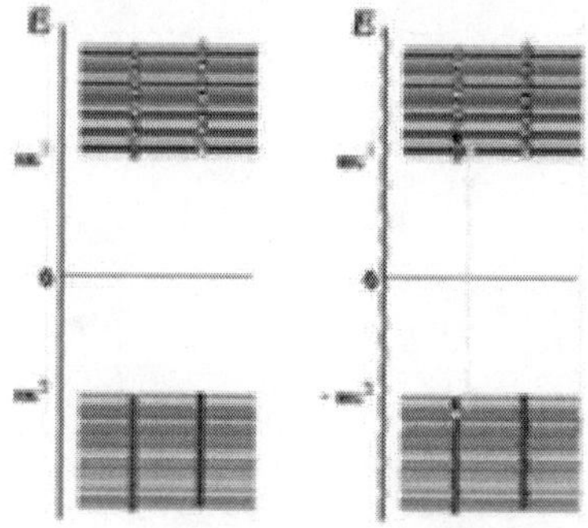

Fig. A Schematic Picture of the Occupied and Empty Levels in the Dirac equation.

The promotion of a particle to an empty level corresponds to the creation of a positron-electron pair, and takes an energy larger than $2mc2$. In figure we have sketched the solutions for the Dirac equation for a free particle. It has a positive energy spectrum starting at *mc*2 (you cannot have a particle at lower energy), but also a negative energy spectrum below-*mc*2. The interpretation of the positive energy states is natural-each state describes a particle moving at an energy above *mc*2. Since we cannot have negative energy states, their interpretation must be very different.

The solution is simple: We assume that in an *empty* vacuum all negative energy states are filled (the "Dirac sea"). Excitations relative to the vacuum can now be obtained by adding particles at positive energies, or creating *holes* at negative energies. Creating a hole takes energy, so the hole states appear at positive energies. They do have opposite charge to the particle states, and thus would correspond to positrons! This shows a great similarity to the behaviour of semiconductors, as you may well know. The situation is explained in figure.

Note that we have ignored the infinite charge of the vacuum (actually, we subtract it away assuming a constant positive background charge.) Removing infinities from calculations is a frequent occurrence in relativistic quantum theory (RQT). Many *unmeasurable* quantities become infinite, and we are only interested in the finite part remaining after removing the infinities. This process is part of what is called renormalisation, which is a systematic procedure to extract finite information from infinite answers!

QED: PHOTON COUPLES TO E^+E^-

We know that electrons and positrons have charge and thus we need to include electrodynamics in the relativistic quantum theory of the electron. That is even more clear when we take into account that an electron and positron can annihilate by emitting two photons (the well-known 511 keV lines), e^+e^-. There is a natural way to describe this coupling, in a so-called Lagrangian approach, which I shall not discuss here. It teaches us that an electron can emit a photon, as indicated figure.

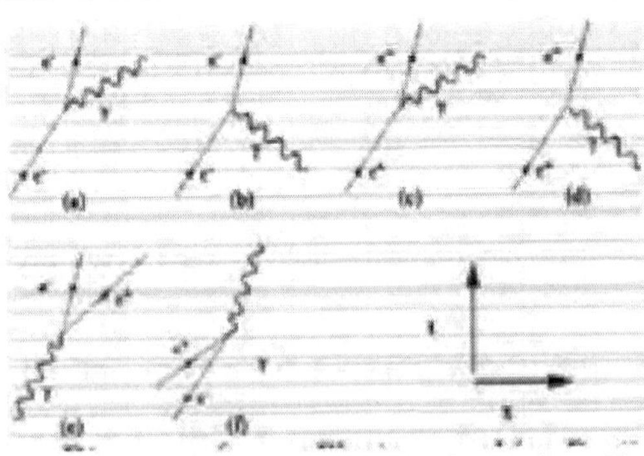

Fig. The Feynman Diagrams for an Electron and/or Positrons Interacting with a Photon. (a) is emission of a photon by an electron, (b) absoption. (c) and (d) are the same diagrams for positrons, and (e) is pair creation, whereas (f) is annihilation.

The diagrams in figure are usually referred to as a Feynman diagrams, and the process depicted in (a) is usually called Bremsstrahlung, the one in (b) annihilation. With such a diagram comes a recipe for calculating it (called the Feynman rule). A key point is that energy and momentum are conserved in all reactions. Let us look at what happens when another nearby electron absorbs the photon, as in figure.

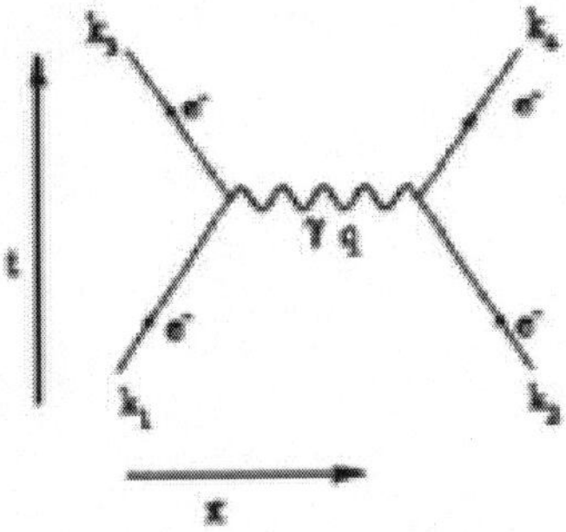

Fig. One of the Feynman Diagrams for an Electron-electron scattering.

- *Of course there are two possibilities*: The left electron can emit the photon to the right one, or absorb one that is emitted by the right one. This is related to the time-ordering of interactions. One of the advantages of feynman diagrams is that both these possibilities are described in one feynman diagram. Thus the time in this diagram should only be interpreted in the sense of the external lines, what are the particles in and out. It is also very economical if we have more and more particles emitted and aborbed.

Since the emitted photon only lives for a short time, $\Delta t = \Delta x / c$, its energy cannot be determined exactly due to the uncertainty relation

$$\Delta E \Delta t \geq \frac{\hbar}{2}.$$

Thus even though the sum of the initial (four) momenta, $k_1 + k_2$ equals the sum of the final ones, $k_3 + k_4$, we find that the photon does *not* have to satisfy.

$$q^2 = E_q^{\ 2} - q^2 = 0.$$

Such a photon is called *virtual* or "off mass-shell", since it does not satisfy the mass-energy relations. This is what gives rise to the Coulomb force.

FLUCTUATIONS OF THE VACUUM

The great problem is in understanding the meaning of virtual particles. Suppose we are studying the vacuum state in QED. We wish to describe this vacuum in terms of the states of no positrons, electrons and photons (the naive vacuum). Since these particles interact we have short-lived states where e^+e^- pairs, and photons, and.... appear for a short while and disappear again. This is also true for real particles: a real electron is a "bare" electron surrounded by a cloud of virtual photons, e^+e^- pairs, etc. A photon can be an e^+e^- pair part of the time, and more of such anomalies.

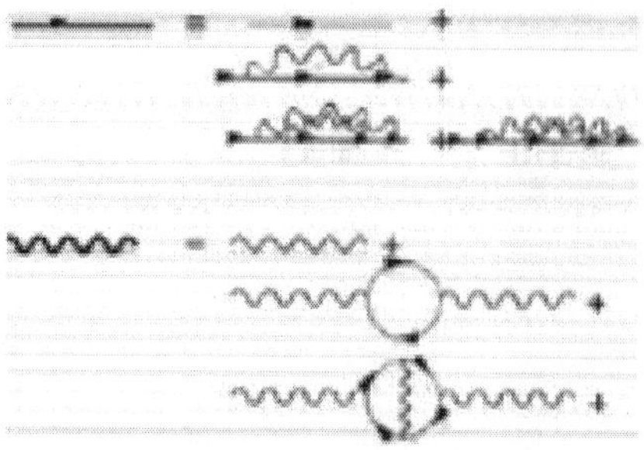

Fig. Some Feynman Diagrams for "Dressed Propagators".

FEYNMAN DIAGRAMS

As I have sketched above, Feynman diagrams can be used to describe what is happening in these processes. These describe the matrix elements, and the actual transition probability is proportional to the square of this matrix elements. One can show that each electron-photon coupling vertex is proportional to e, and thus in the square each vertex gives a factor e^2. Actually by drawing time in the vertical direction and space in the horizontal (schematically, of course), we see that the the two possible couplings of the photon to matter-Bremsstrahlung and pair creation are one and the same process. Still it helps to distinguish. Note that at each vertex charge is conserved as well as momentum! This can actually be combined into a dimensionless quantity,

$$\alpha = \frac{e^2}{4\pi\varepsilon_0 \hbar c} \approx \frac{1}{137}$$

We should expand in rather than $e2$ since expansion parameters, being "unphysical" can not have dimensions. In other words in order to carry through this mathematical concept the natural scale of a diagram is set by the power of it carries. Due to the smallness of we normally consider only the diagrams with as few vertices as possible. Let me list the two diagrams for electron- positron scattering, both proportional to α^2, as given in figure.

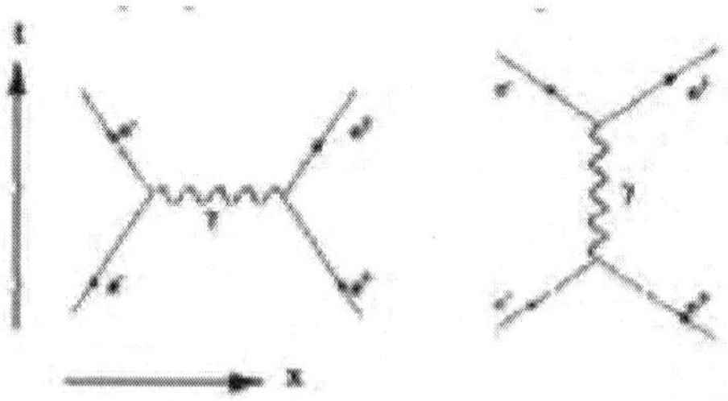

Fig. The two Feynman Diagrams for an Electron-positron scattering.

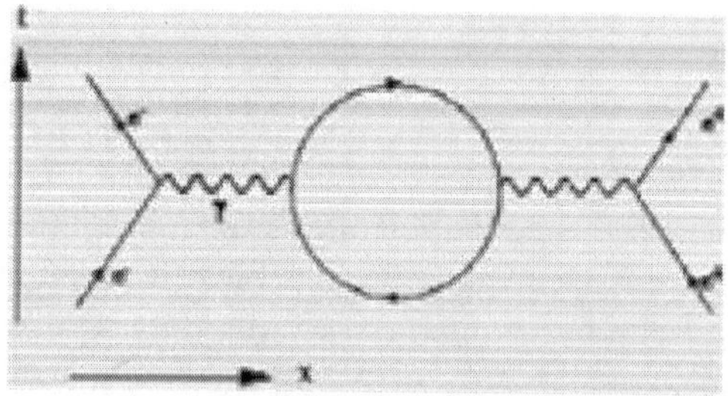

Fig. A Higher Order Diagram for Electron-positron Scattering.

The sum of all diagrams contributing to a given process is called the perturbation series.

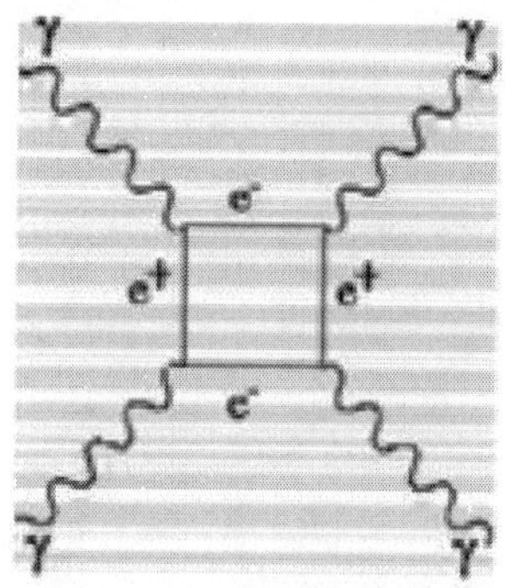

Fig. The lowest diagram for photon-photon scattering.

INFINITIES AND RENORMALISATION

One of the key features missing in the discussion above is the fact that all the pictures I have drawn are infinite-somewhat of a severe blow. The key point is to understand that this is not a problem, but has to do with a misinterpretation of the series. When we introduce a and *e* in our theory these we use the measured value of the charge of an electron-which is a solution to the full theory, not to the artificial problem with all vacuum fluctuations turned of. What it means is that we should try to express all our answers in physically sensible (measurable) quantities. Renormalisation is the mathematical procedure that does this. A theory (such as QED) is called renormalisable if we can make all expressions finite by reexpressing them in a finite number of physical parameters.

WAVE EQUATIONS AND WAVEPACKETS

De Broglie's doctoral thesis, defended at the end of 1924, created a lot of excitement in European physics circles. Shortly after it was published in the fall of 1925 Pieter Debye, a theorist in Zurich, suggested to Erwin Schrodinger that he give a seminar on de Broglie's work. Schrodinger gave a polished presentation, but at the end Debye remarked that he considered the whole theory rather childish: why should a wave confine itself to a circle in space?

It wasn't as if the circle was a waving circular string, real waves in space diffracted and diffused, in fact they obeyed three-dimensional wave equations, and that was what was needed. This was a direct challenge to Schrodinger, who spent some weeks in the Swiss mountains working on the problem, and constructing his equation. There is no rigourous derivation of Schrodinger's equation from previously established theory, but it can be made very plausible by thinking about the connection between light waves and photons, and construction an analogous structure for de Broglie's waves and electrons.

THE WAVE EQUATION TELL US ABOUT THE PHOTON

We know from the photoelectric effect and Compton scattering that the photon energy and momentum are related to the frequency and wavelength of the light by

$$E = hv = \hbar\omega$$

$$p = \frac{h}{\lambda} = \hbar k$$

Notice, then, that the wave equation tells us that $\omega = ck$ and hence $E = cp$.

To put it another way, if we think of $e^{i(kx-wt)}$ as describing a particle (photon) it would be more natural to write the plane wave as

$$\vec{E}_0 e^{\frac{i}{k}(px-Et)}$$

that is, in terms of the energy and momentum of the particle.

In these terms, applying the (Maxwell) wave equation operator to the plane wave yields.

$$\left(\frac{\partial^2}{\partial x^2} - \frac{1}{c^2}\frac{\partial^2}{\partial t^2}\right)\vec{E}_0 e^{\frac{i}{\hbar}(px-Et)} = \left(p^2 - \frac{E^2}{c^2}\right)\vec{E}_0 e^{\frac{i}{\hbar}(px-Et)} = 0$$

or,

$$E^2 = c^2 p^2.$$

The wave equation operator applied to the plane wave describing the particle propagation yields the energy-momentum relationship for the particle.

CONSTRUCTING A WAVE EQUATION OPERATOR

The wave equation operator from the photon case (zero rest mass) to a particle having rest mass $m0$. We need a wave equation operator that, when it operates on a plane wave, yields

$$E^2 = c^2 p^2 + m_0^2 c^4$$

Writing the plane wave function,

$$\phi(x,t) = Ae^{\frac{i}{\hbar}(pk-Et)}$$

where is a contant, we find we can get $E^2 = c^2 p^2 + m_0^2 c^4$ by adding a constant (mass) term to the differentiation terms in the wave operator:

$$\left(\frac{\partial^2}{\partial x^2} - \frac{1}{c^2}\frac{\partial^2}{\partial t^2} - \frac{m_0^2 c^2}{\hbar^2}\right) Ae^{\frac{i}{\hbar}(px-Et)} = \frac{1}{\hbar^2}\left(p^2 - \frac{E^2}{c^2} + m_0^2 c^2\right) Ae^{\frac{i}{\hbar}(px-Et)=0}$$

This wave equation is called the *Klein-Gordon* equation and correctly describes the propagation of relativistic particles of mass m_0. However, it's a bit inconvenient for non-relativistic particles, like the electron in the hydrogen atom, just as $E^2 = m_0^2 c^4 + c^2$ is less useful than $E = p^2/2m$ for this case.

A NON-RELATIVISTIC WAVE EQUATION

Continuing along the same lines, let us assume that a non-relativistic electron in free space (no potentials, so no forces) is described by a plane wave:

$$\psi(x,t) = Ae^{\frac{i}{\hbar}(px-Et)}$$

We need to construct a wave equation operator which, applied to this wave function, just gives us the ordinary non-relativistic energy-momentum relationship, $E = p^2/2m$. The $p2$ obviously comes as usual from differentiating twice with respect to x, but the only way we can get E is by having a *single* differentiation with respect to time, so this looks different from previous wave equations:

$$i\hbar\frac{\partial\psi(x,t)}{\partial t} = -\frac{\hbar^2}{2m}\frac{\partial^2\psi(x,t)}{\partial x^2}.$$

This is *Schrodinger's equation* for a free particle. It is easy to check that if y (x, t) has the plane wave form given above, the condition for it to be a solution of this wave equation is just $E = p^2/2m$. Notice one remarkable feature of the above equation the i on the left means that y *cannot* be a real function.

VARYING POTENTIAL AFFECT A DE BROGLIE WAVE

The effect of a potential on a de Broglie wave was considered by Sommerfeld in an attempt to generalize the rather restrictive conditions in Bohr's model of the atom. Since the electron was orbiting in an inverse square force, just like the planets around the sun, Sommerfeld couldn't understand why Bohr's atom had only circular orbits, no Kepler-like ellipses. (Recall that all the observed spectral lines of hydrogen were accounted for by energy differences between *circular* orbits.)

De Broglie's analysis of the allowed circular orbits can be formulated by assuming at some instant in time the spatial variation of the wave function on going around the orbit includes a phase term of the form *eipq* / h , where here the parameter q measures distance around the orbit.

Now for an acceptable wave function, the total phase change on going around the orbit must be 2*n*p , where n is an integer. For the usual Bohr circular orbit, p is constant on going around, q changes by $2\pi r$, where r is the radius of the orbit, giving,

$$\frac{1}{\hbar}p2\pi r = 2n\pi, so pr = n\hbar,$$

the usual angular momentum quantization. What Sommerfeld did was to consider a general Kepler ellipse orbit, and visualize the wave going around such an orbit. Assuming the usual relationship $p = h/\lambda$ the wavelength will vary as the particle moves around the orbit, being shortest where the particle moves fastest, at its closest approach to the nucleus. Nevertheless, the phase change on moving a short distance Δq should still be $p\Delta q/\hbar$, and requiring the wave function to link up smoothly on going once around the orbit gives,

$$\oint pdq = nh.$$

Thus only certain elliptical orbits are allowed. The mathematics is non-trivial, but it turns out that *every allowed elliptical orbit has the same energy as one of theallowed circular orbits*.

That is why Bohr's theory gave all the energy levels. Actually, this whole analysis is old fashioned (it's called the "old quantum theory") but we've gone over it to introduce the idea of *a wave with variable wavelength, changing with the momentum as the particle moves through a varying potential*. The reader may well be wondering at this point why it is at all useful to visualize a real wave going round an orbit, when we have stated that any solution of Schrodinger's equation is necessarily a complex function. Including those corresponding to Bohr's energy levels, in which the complex nature of the wave function only appears in a time varying phase factor $e^{-iEt/\hbar}$. We should also add that if the spatial dependence is a real function, such as sin *kx*, it represents a standing wave, not a particle circling in one direction, which would be e^{ikx}, or e^{-iEx} / h Bearing all this in mind, it is still often instructive to sketch real wave functions, especially for one-dimensional problems.

SCHRODINGER'S EQUATION FOR A PARTICLE IN A POTENTIAL

In quantum mechanics and scattering theory, the one dimensional step potential is an idealized system used to model incident, reflected and trans mitted matter waves. The problem cons ists of solving the time- independent Schrödinger equation for a particle with a step-like potential in one dimension. Typically, the potential is modelled as a Heaviside step function.

Let us consider first the one-dimensional situation of a particle going in the *x*-direction subject to a "roller coaster" potential. What do we expect the wave function to look like? We would expect the wavelength to be shortest where the potential is lowest, in the valleys, because that's where the particle is going fastest maximum momentum.

With a non-zero potential present, the energy-momentum relationship for the particle becomes the energy equation,

$$E = \frac{p^2}{2m} + V(x)$$

We need to construct a wave equation which leads naturally to this relationship. In contrast to the free particle cases, the relevant wave function here will no longer be a plane wave, since the wavelength varies with the potential. However, at a given x, the momentum is determined by the "local wavelength", that is,

$$p = -i\hbar\frac{\partial\psi}{\partial x}$$

It follows that the appropriate wave equation is:

$$i\hbar\frac{\partial\psi(x,t)}{\partial t} = -\frac{\hbar^2}{2m}\frac{\partial^2}{2m}\frac{\partial^2\psi(x,t)}{\partial x^2} + V(x)\psi(x,t).$$

This is the standard one-dimensional Schrodinger equation.

In three dimensions, the argument is precisely analogous. The only difference is that the square of the momentum is now a sum of three squared components, for the x, y and z directions, so,

$$\frac{\partial^2}{\partial x^2}\ becomes\ \frac{\partial^2}{\partial x^2} + \frac{\partial^2}{\partial y^2} + \frac{\partial^2}{\partial z^2} = \nabla^2$$

so now,

$$i\hbar\frac{\partial\psi(x,y,z,t)}{\partial t} = -\frac{\hbar^2}{2m}\nabla^2\psi(x,y,t) + V(x,y,z)\psi(x,y,z,t).$$

This is the complete Schrodinger equation. So far, of course, it is based on plausibility arguments and hand-waving. Why should anyone believe that it really describes an electron wave? Schrodinger's test of his equation was the hydrogen atom. He looked for Bohr's "stationary states": states in which the electron was localized somewhere near the proton, and having a definite energy. The time dependence would be the same as for a plane wave of definite energy, $e^{-iEx/h}$ the spatial dependence would be a time-independent function decreasing rapidly at large distances from the proton. That is, he took,

$$\psi(x,y,z,t) = e^{-ZEt/\hbar}\psi(x,y,z)$$

He took advantage of the spherical symmetry by re-expressing the spatial wave function in spherical polar coordinates, and found his equation became a standard differential equation solved in the nineteenth century. The solution gave the shape of possible wave functions, and also allowed values of energy and angular momentum. These values were exactly the same as Bohr's (except that the lowest allowed state in the new theory had zero angular momentum): impressive evidence that the new theory was correct.

PRESENT CONSERVATION

When Schrodinger published this result in 1926, he also wrote down the complex conjugate equation, and proved that taking them together it was not difficult to deduce a continuity equation:

$$\frac{\partial \rho}{\partial t} + div\vec{j} = 0$$

$$where\ \rho = \psi^* \psi = |\psi|^2$$

$$and\ \vec{j} = \frac{\hbar}{2mi}\left(\psi^* \vec{\Delta}\psi - \psi \vec{\Delta}\psi^*\right).$$

Schrodinger believed the above continuity equations represented the conservation of electric charge, and had no further significance. He thought that after all his own equation showed the electron to be just a smooth classical wave at the deepest level. In fact, he succeeded in solving the three-dimensional equation with a Coulomb potential and he found the Bohr energy levels of the hydrogen atom. Obviously, he was on the right track! This classical revival approach, however, couldn't deal with the unpredictability of quantum mechanics, such as where a single photon or electron would land in a two-slit diffraction pattern. The truth is, Schrodinger didn't understand his own equation.

Another physicist, Max Born, published a paper a few days after Schrodinger's in which he suggested that

$$\left|\psi(x,y,z,t)^2\right| dxdydz$$

was the *relative probability* of finding the electron in a small volume *dxdydz* at (*x,y,z*) at time *t*. This interpretation was based directly on the analogy with light waves and photons, and has turned out to be correct. *Notation note*: y is called the "*amplitude*" or sometimes the "*probability amplitude*".

PHOTONS AND ELECTRONS

In physics, a photon is an elementary particle, the quantum of light and all other forms of electromagnetic radiation, and the force carrier for the electromagnetic force. The effects of this force are easily observable at both the microscopic and macroscopic level, because the photon has no rest mass; this allows for interactions at long distances. Like all elementary particles, photons are currently best explained by quantum mechanics and exhibit wave– particle duality, exhibiting properties of both waves and particles. For example, a single photon may be refracted by a lens or exhibit wave interference with itself, but also act as a particle giving a definite result when its position is measured.

The electron is a subatomic particle with a negative elementary electric charge. It has no known components or substructure; in other words, it is generally thought to be an elementary particle. An electron has a mass that is approximately 1/1836 that of the proton. The intrinsic angular momentum (spin) of the electron is a half-integer value in units of *h*, which means that it is a fermion.

The antiparticle of the electron is called the positron; it is identical to the electron except that it carries electrical and other charges of the opposite sign. When an electron collides with a positron, both particles may be totally annihilated, producing gamma ray photons. Electrons, which belong to the first generation of the lepton particle family, participate in gravitational, electromagnetic and weak interactions.

Electrons, like all matter, have quantum mechanical properties of both particles and waves, so they can collide with other particles and can be diffracted like light. However, this duality is best demonstrated in experiments with electrons, due to their tiny mass. Since an electron is a fermion, no two electrons can occupy the same quantum state, in accordance with the Pauli exclusion principle.

We have seen that electrons and photons behave in a very similar fashion both exhibit diffraction effects, as in the double slit experiment, both have particle like or quantumbehaviour. We now have a framework for understanding photons we first figure out how the electromagnetic wave propagates, using Maxwell's equations, that is, we find E as a function of x, y, z, t. Having evaluated $E(x, y, z, t)$, the probability of finding a photon in a given small volume of space $dxdydz$, at time t, is proportional to $|E(x,y,z,t)|^2 dxdydz$, the energy density. Born assumed that Schrodinger's wave function for the electron corresponded to the electromagnetic wave for the photon in the sense that the square of the modulus of the Schrodinger wave amplitude at a point was the relative probability density for finding the electron at that point.

So the routine is the same: for given boundary conditions and a given potential, Schrodinger's differential equation can be solved and the wave function

$$\psi(x, y, z, t)$$

valuated. Then,

$$|\psi(x, y, z, t)|^2 \, dxdydz$$

gives the relative probability of finding the electron at (x,y,z) at time t.

Notice, though, that this interpretation of the wave function is *not* essential in finding the allowed energy levels in a given potential, such as the Bohr orbit energies, which Schrodinger derived before the physical significance of his wave function was understood.

Roller Coaster Potential

An electron travelling along a roller coaster potential, the local wavelength is related to the momentum of the electron as it passes that point. Perhaps slightly less obvious is that the *amplitude* of the wave varies: it will be largest at the tops of the hills (provided the particle has enough energy to get there) because that's where the particle is moving slowest, and therefore is most likely to be found.

Suppose following de Broglie we write down the relation between the "particle properties" of the electron and its "wave properties":

$$\frac{1}{2}mv^2 = E = hf, mv = h/\lambda$$

It would seem that we can immediately figure out the speed of the wave, just using l*f* = *c* ' say.

We find:

$$\lambda f = \frac{h}{mv} \cdot \frac{\frac{1}{2}mv^2}{h} = \frac{1}{2}v.$$

So the speed of the wave seems to be only half the speed of the electron! How could they stay together? What's wrong with this calculation?

To answer this question, it is necessary to think a little more carefully about the wave function corresponding to an electron traveling through a cathode ray tube, say. The electron leaves the cathode, shoots through the vacuum, and impinges on the screen. At an intermediate point in this process, it is moving through the vacuum and the wave function must be non-zero over some volume, but zero in the places the electron has not possibly reached yet, and zero in the places it has definitely left.

However, if the electron has a precise energy, say exactly a thousand electron volts, it also has a precise momentum. This necessarily implies that the wave has a precise wavelength. But the only wave with a *precise* wavelength λ has the form,

$$\psi(x,t) = Ae^{i(kx-\omega t)}$$

where $k = 2\pi/\lambda$ *and* $\omega = 2\pi f$ The problem is that this plane sine wave extends to infinity in both spatial directions, so cannot represent a particle whose wave function is non-zero in a limited region of space. Therefore, to represent a localized particle, we must superpose waves having different wavelengths. Now, the waves representing electrons, *unlike the light waves representing photons*, travel at different speeds for different energies.

Any intuition gained by thinking about superposing light waves of different wave lengths can be *misleading* if applied to electron waves. Fortunately, there are many examples in nature of waves whose speed depends on wavelength. A simple example is water waves on the ocean. We consider waves having a wavelength much shorter than the depth of the ocean . What is the ω, k relationship for these waves? We knowit's not $\omega = Ck$, with a constant *C*, because waves of different wavelengths move at different speeds.

In fact, it's easy to figure out the ω, k relationship, known as the *dispersion relation*, for these waves from a simple dimensional argument. What physical parameters can the wave frequency depend on? Obviously, the wave length λ . We w ill use $k = 2\pi/\lambda$ as our variable. *k* has dimensions L^{-1}.

These waves are driven by gravity, so g, with dimensions LT^{-2}, is relevant. Actually, that's all. For ocean waves, surface tension is certainly negligible, as is the air density, and the water's viscosity. You might think the density of the water matters, but these waves are rather like a pendulum, in that they are driven by gravity, so increasing the density would increase both force and inertial mass by the same amount.

For these deepwater waves, then, dimensional analysis immediately gives:

$$\omega^2 = Cgk$$

where C is some dimensionless constant we cannot fix by dimensional argument, but actually it turns out to be 1.

ATTITUDE OF SUPERPOSITION

To return momentarily to the electron traveling through a vacuum, it is clear physically that it must have a wave function that goes to zero far away in either direction. A localized wave function of this type is called a "wavepacket". We shall discover that a wavepacket can be constructed by adding plane waves together. Now, the plane waves we add together will individually be solutions of the Schrodinger equation.

But does it follow that the *sum* of such solutions of the Schrodinger equation is itself a solution to the equation?

The answer is yes in other words, the Schrodinger equation:

$$i\hbar\frac{\partial\psi(x,y,z,t)}{\partial t} = -\frac{\hbar^2}{2m}\nabla^2\psi(x,y,z,t) + V(x,y,z)\psi(x,y,z,t)$$

is a *linear* equation, that is to say, if y1 (x, y, z,t) , y2 (x, y, z, t) are both solutions of the equation, then so is,

$$\psi(x,y,z,t) = c_1\psi_1(x,y,z,t) + c_1\psi_2(x,y,z,t)$$

where c^1 and c^2 are arbitrary constants, as is easy to check. This is called the *Principle of Superposition*.

The essential point is that in Schrodinger's equation every term contains a factor y, but no term contains a factor ψ^2 (or a higher power). That's what is meant by a "linear" equation. If the equation *did* contain a constant term, or a term including ψ^2, superposition wouldn't work the sum of two solutions to the equation would not itself be a solution to the equation. In fact, we have been assuming this linearity all along: when we analyse interference and diffraction of waves, we just add the two wave amplitudes at each spot. For the double slit, we take it that if the wave radiating from one slit satisfies the wave equation, then adding the two waves together will give a new wave which also satisfies the equation.

❐

4.

Chemical Bonding

CHEMICAL BOND

Chemical bond referrs to the forces holding atoms together to form molecules and solids. This force is of electric nature, and the attraction between electrons of one atom to the nucleus of another atom contribute to what is known as chemical bonds. Although electrons of one atom repell electrons of another, but the repulsion is relatively small. So is the repulsion between atomic nuclei. Various theories regarding chemical bond have been proposed over the past 300 years, during which our interpretation of the world has also changed.

Some old concepts such as Lewis dot structure and valency are still rather useful in our understanding of the chemical properties of atoms and molecules, and new concepts involving quantum mechanics of chemical bonding interpret modern observations very well.

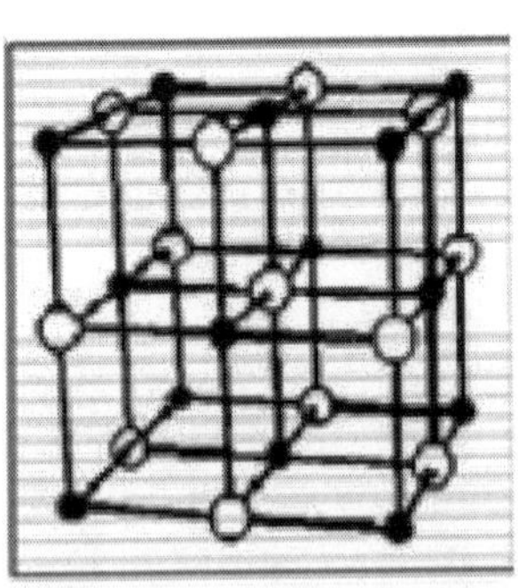

While reading, you learn new concepts such as bondlength, bond energy, bond order, covalent bond, ionic bond, polar and non-polar bond etc. These concepts help you understand the material world at the molecular level.

Chemical bonds between identical atoms such as those in H_2, N_2, and O_2 are called covalent bonds, in which the bonding electrons are shared. In ionic compounds, such as NaCl, the ions gather and arrange in a systematic fashion to form a solid. The arrangement of (blue) Na^+ and (green) Cl^- ions in a solid is shown in on the right here. The attraction force between ions are called ionic bonding. Matals such as sodium, copper, gold, iron etc. have special properties such as being good electric conductors. Electrons in these solids move freely throughout the entire solid, and the forces holding atoms together are called metallic bonds. To some extend, metals are ions submerged is electrons.

BASIC OF CHEMICAL BOND

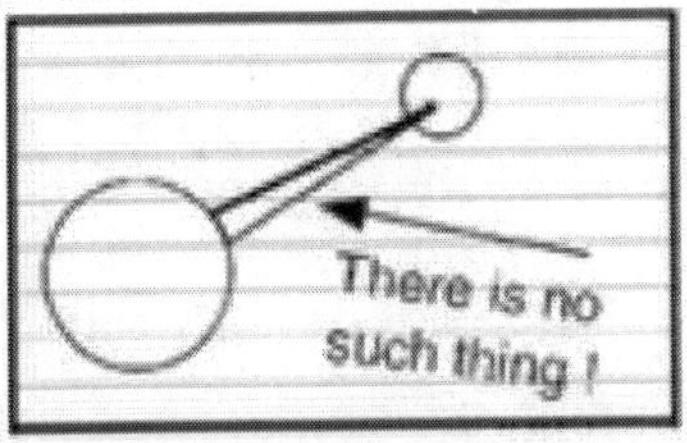

You probably learned some time ago that chemical bonds are what hold atoms together to form the more complicated aggregates that we know as molecules and extended solids. Chemists talk about bonds all the time, and draw pictures of them as lines joining atom symbols. Teachers often identify them as the little sticks that connect the spheres that represent atoms in a plastic molecular model. So it's not surprising that we sometimes tend to think of chemical bonds as "things". But no one has ever *seen* a chemical bond, and there is no reason to believe that they really even exist as physical objects.

So although the "chemical bond" is no more than a convenient fiction, *chemical bonding*, which leads to the near-infinity of substances, lies at the very core of chemistry.

The forces that hold bonded atoms together are basically just the same kinds of electrostatic attractions that bind the electrons of an atom to its positively-charged nucleus;

This is the most important fact about chemical bonding that you should know, but it is not of itself a workable *theory* of bonding because it does not describe the conditions under which bonding occurs, nor does it make useful predictions about the properties of the bonded atoms.

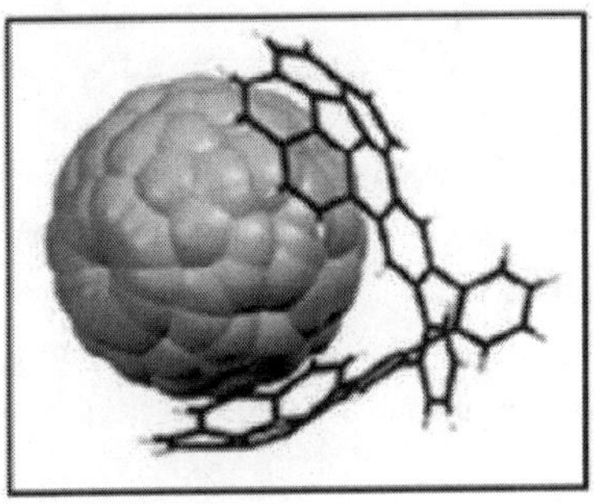

CHEMICAL BOND CONCEPTS

Chemical bonds form to lower the energy of the system, the components of the system become more stable through the formation of bonds. Inert gases (Noble gases) are stable due to a full valence shell of electrons (ns^2, np^6). Bonds form to enable elements to attain this 'noble gas configuration'. Ionic bonds involve the transfer of electrons from a metal to a non-metal, forming positive and negative ions that are attracted to each other. Covalent bonds ~8 formed by overlap (or combination) of atomic orbitals of each element to form a molecular orbital. Metallic bonds can be seen as each metal donating electron(s) to a common "sea" of electrons which are shared by all the ions within the solid.

Hydrogen bonding: The oxygen-hydrogen bond is polar, oxygen being more electronegative element. The molecule is therefore polar (the molecule is not linear but has a bent, V, shape). This is extenuated by the two lone pairs of electrons on the oxygen atoms. One end of the molecule is partially negative while the two hydrogen atoms become partially positive. The molecules of water are attracted to one another, with the slightly positive hydrogens attracted to the negative 'ends' (the oxygen of other water molecules). This intermolecular attraction is termed hydrogen bonding and acts almost like a glue holding the molecules of water together.

If we hook hydrogen up to an atom that is very good at attracting electrons (like N, O, or F), the hydrogen end of the bond becomes very positively charged and the other atom becomes negatively charged (polar). Remember that hydrogen is the smallest atom on the periodic table, it is possible for two molecules to get very close together. The combination of high polarity and close approach result in the interaction being particularly strong. Coulombic forces: attractive (or repulsive) forces between charged particles. The force of attraction between No opposite charges is proportional to the magnitude of their charges divided by the square of the distance between them.

Hydrogen bonding produces adhesion, cohesion, surface tension, anal capillary action. In polar molecules, the presence of oppositely charged ends produces forces of attraction between these molecules that are greater than the forces of attraction between similar non-polar molecules. These forces affect the properties of the polar substances. Some of the effects include an increase in the boiling and melting points and higher heats of vapourization and fusion. These greater intermolecular forces also are likely to account for the lower vapor pressure of substances having polar molecules. Also, polar molecules have a greater attraction for ions.

It is the polarity of its molecules that makes water a good solvent. Many ionic compounds readily dissolve in water. The H-O bonds in water are polar because O atoms have a higher electronegativity than H atoms. If the water molecule were linear, it would be non-polar because the bonds in a linear water molecule would be symmetrically placed. The lack of symmetry in electron distribution gives a partial negative charge to the end of the molecule at the greatest distance from both hydrogen atoms. At the opposite end, there is a partial positive charge.

The hydrogen atom in a hydrogen bond is in effect bonded to two atoms, more weakly to one than the other. It is bonded covalently to an atom within its own m olec ule a nd, through the h ydrogen bond, to a n atom in a neighbouring molecule. The strength of the hydrogen bond increases with the degree of electronegativity of the atom bonded to the hydrogen. The strength decreases with an increase in the size of the bonded atom. For example, nitrogen and chlorine atoms have nearly the same electronegativies.

However, the hydrogen bond between hydrogen in one molecule and nitrogen in an adjacent one is much stronger than the bond between hydrogen and an adjacent chlorine atom. This is because nitrogen atoms are much smaller than chlorine atoms. The negative charge of the electrons in the nitrogen atom is concentrated into a smaller volume, and it therefore exerts a greater attraction for the proton of the hydrogen atom in a neighbouring molecule. In fact, the bond between the hydrogen atom and an adjacent chlorine atom is not considered to be a hydrogen bond because it does not have a sufficiently great force of attraction. In any molecule where a hydrogen atom is bonded to one of the small, highly electronegative atoms F, O, or N, hydrogen bonding is likely to occur.

Hydrogen bonding also explains why some substances have unexpectedly low vapour pressures, high heats of vaporization, and high melting points. In order for vaporization or melting to take place, molecules must be separated. An input of energy is required to break hydrogen bonds between molecules and thus break down the larger clusters of molecules into separate molecules. As with the boiling point, the melting point of water is abnormally high when compared with the melting points of the hydrogen compounds of the other elements from Group 16. These substances are chemically similar but have no appreciable hydrogen bonding.

Hydrogen bonding has an effect on the crystal structure of ice. X ray studies show that the three dimensional structure caused by hydrogen bonding gives the ice crystals a crystalline stricture with many hexagonal openings. This open structure accounts for the low density of ice. When ice is melted, hydrogen bonds are broken. Then the open structure is destroyed, and molecules move closer together Therefore, the liquid phase of water has a greater density than the solid phase. For most substances, the solid phase has a greater density than the liquid phase..

Hydrogen bonds are mainly responsible for the coiled shape of protein molecules. Hydrogen bonds are found in nucleic acids DNA and RNA. The bonds hold together the double helix structure in the DNA.

METALLIC BONDING

Most metals have only one or two valence electrons and low ionization energies. Their valence electrons are not tightly bound to the atom, but seem to move easily from one atom to another. They can be considered a part of the whole metal crystal. These mobile electrons exert an attractive force ors t he positive ions, helping to fix their positions.

The ease with which the valence electrons move within the crystal distinguishes the metallic bond from the ionic or covalent bond:

- Metals are good conductors of heat and electricity because of the mobility of their valence electrons.
- The binding action of the electrons is the basis for the hardness of metals. Softer metals have weaker binding forces.
- The high luster of metals is the result of the uniform way in which the valence electrons absorb and re emit light energy that strikes them.
- The malleability (ability to be flattened into thin foil without breaking), ductility (ability to be stretched into a thin wire without breaking), and sectility (ability to be cut into sections without shattering) result from the fairly uniform attraction between the electrons and the ions. The ions can change position, or 'flow' in the 'sea' of valence electrons. Metals can be flattened or stretched into a wire because the electrons and ions can move into other positions without breaking up the essential structure. The attraction between electrons and ions continues even while forces are applied that change the shape of the metal.

Various concepts or theories have been proposed to explain the formation of compounds. In particular, chemical bonds were proposed to explain why and how one element reacted with another element. In 1852, E. Frankland proposed the concept of valence. He suggested that each element formed compounds with definite amounts of other elements due to a valence connection. Each element has a definite number of valance.

Five years later, F.A. Kekule and others proposed a valence of 4 for carbon. Lines were used to represent valance, and this helped the development of organic c hemistry. Th e structure of ben zene w as often q uoted as an achievement in this development. More than 10 years later, J.H. van't Hoff and le Bel proposed thetetrahedral arrangement for the four valences around the c arbon . Th ese th eory h elped c hem ists to des cribe m an y orga nic compounds. In the mean time,chemical bonds were thaught to be electric nature. Since electrons have not been discovered as the negative charge carriers, they were thought to be involved in chemical bonds.

Following the discoveries of electrons by J.J. Thomson and R. A. Millikan, G.N. Lewis proposed to use dots to represent valence electrons. His dots made the valence electrons visible to chemists, and he has been credited as the originator of modern bonding theory. He published a book, in 1923, called Valence and the Structure of Atoms and Molecules.

X-ray diffractions by crystal allow us to calculate details of bondlength and bond angles. Using computers, we are able to generate images of molecules from the data provided by X-ray diffraction studies. These data prompted Linus Pauling to look at The Nature of Chemical Bond, a book that introduced many new concepts such as the resonance, electronegativity, ionic bond, and covalent bond.

In England, N.V. Sidgwick and H.E. Powell paid their attention to the lone pairs in a molecule. They developed the valence bond theory, the VSEPR (Valence Shell Electron Pair Repulsion) theory. The application of quantum theory to chemical bonding gave birth to a molecular orbital theory.

BASIC BONDING CONCEPTS

The most basic concept in all of chemistry is the chemical bond. The chemical bond is essentially the sharing of electrons between two atoms, a sharing which holds or bonds the atoms together. In order to understand the chemical bond, atoms and electrons must first be understood.

ELECTRON ORBITALS

Atoms have three components: protons, neutrons, and electrons. Protons have a positive charge of +1, neutrons have 0 charge, and electrons have a negative charge of -1. The protons and neutrons occupy the center of the atom as a piece of solid matter called the nucleus. The electrons exist in "orbitals" surrounding the nucleus. In the olden times, it was believed that the electrons orbited the nucleus in the same way that planets orbit the sun. This is not exactly true, but it explains the origin of the term orbital to describe the location of the electron.

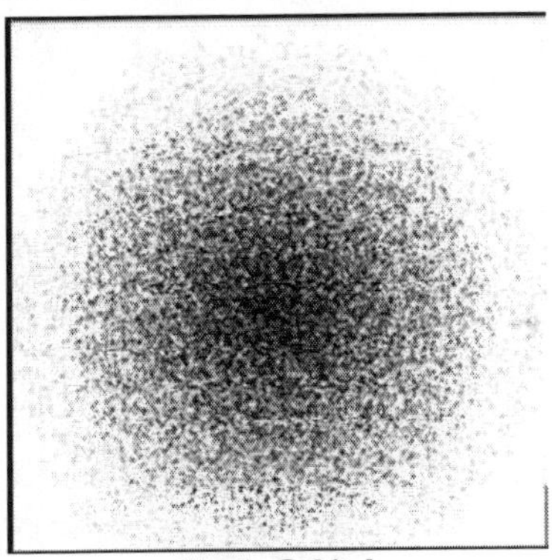

Fig. 1s Orbital

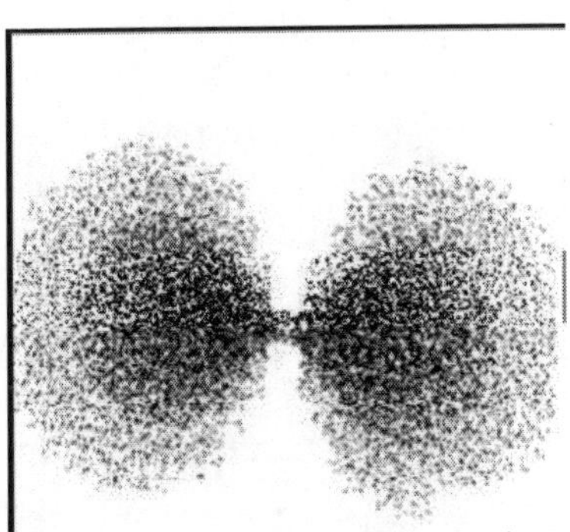

Fig. 2p Orbital

In reality, it is impossible to tell the precise trajectory of an electron in the way that the path of a planet can be followed. The best that can be achieved is to describe the probability of locating the electron in a region of space. The distribution of this probability over space is called an orbital. Let us treat the simplest case, in which the nucleus is surrounded by just one electron (for example, the hydrogen atom). In this case, the probability of finding an electron in its lowest energy, or most stable, state is distributed in a spherically symmetric way around the nucleus. The probability of finding the electron is highest at the nucleus and decreases as the distance from the nucleus increases (Figure).

The density of dots in a region represents to the probability of locating the electron there. This lowest energy, spherically symmetric orbital is called the 1s orbital. As stated, in this orbital, the highest probability of finding the electron at a point in space occurs at the nucleus. Such a location may seem counterintuitive, since the electron is known to generally be located some dis ta nce from th e nucleus, not direc tly a t it. How c an th is apparen t contradiction be explained?

Although the probability of finding the electron at a point is greatest at the nucleus, the volume of space occupied by the nucleus is extremely small compared to the rest of the atom. The probability of finding the electron at a point decreases farther from the nucleus, but the volume in which it can be found becomes much, much larger. So the electron is more likely to be found some distance from the nucleus.

It is the product of the volume in which the electron can be found and the probability of finding it that gives the probability of finding the electron as a function of distance from the nucleus. The radial distribution function, which is a product of the probability and the volume element, gives the probability of finding the electron in a spherical shell some distance r from the nucleus. It is clear that the electron is most likely to be found in a shell some distance from the nucleus rather than directly at the nucleus. In fact, since the volume element is almost zero directly at the nucleus, the electron is very unlikely to be found there. The 1s orbital is the lowest energy orbital that an electron can occupy, but several higher energy orbitals are significant in organic chemistry. The next lowest energy orbital that an electron can occupy is the 2s orbital, which looks much like the 1s orbital except that the electron is more likely to be found farther from the nucleus.

The third lowest energy orbital is called the 2p orbital. The major and highly important difference between a p orbital and an s orbital is that the p orbital is not spherically symmetric . Instead, the p orbital is oriented along a specific axis in space. There are three p orbitals, oriented along the x, y, and z axes.

So far, our discussion of orbitals has been mathematically exact. The images of s and p orbitals depict the actual distribution over space of the probability of finding an electron in atoms with one electron. However, after a second electron is added, the problem cannot be solved exactly due to the electron-electron repulsion. Instead, approximations have to be made. Fortunately, using the exact orbitals of the one-electron atom as a basis for the orbitals in many-electron atoms is an excellent approximation. For example, in carbon, the energy of the electron is correct to within 1 per cent using this approximation. Thus, orbitals in many-electron atoms look essentially like the one-electron atom orbitals that have already been described.

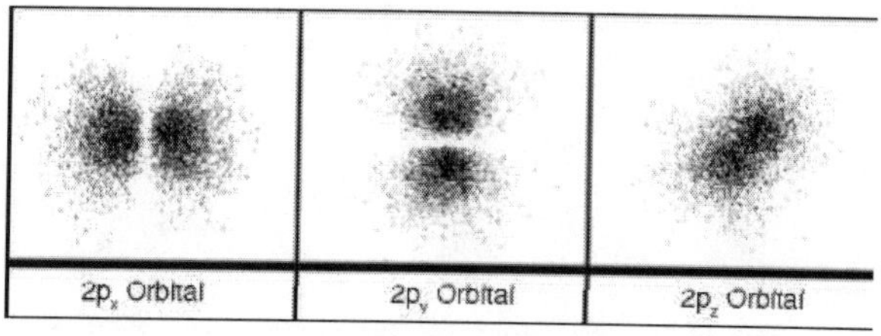

Fig. The Three 2p Orbitals

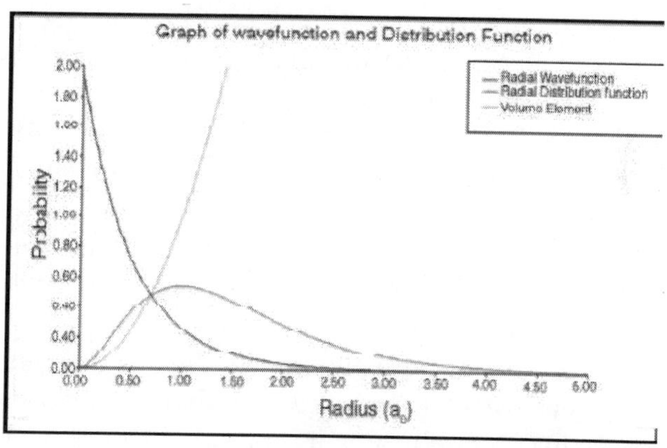

Fig. Graph of Probability Density, Radial Distribution Function, and Volume as a Function of Radius for the 1s Electron

CARBON–CARBON BOND

A carbon–carbon bond is a covalent bond between two carbon atoms. The most common form is the single bond: a bond composed of two electrons, one from each of the two atoms. The carbon–carbon single bond is a sigma bond and is said to be formed between one hybridized orbital from each of the carbon atoms. In ethane, the orbitals are sp^3 hybridized orbitals, but single bonds formed between carbon atoms with other hybridisations do occur (*e.g.* sp^2 to sp^2). In fact, the carbon atoms in the single bond need not be of the same hybridisation. Carbon atoms can also form double bonds called alkenes or triple bonds called alkynes. A double bond is formed with an sp^2 hybridized orbital and a p-orbital that isn't involved in the hybridization. A triple bond is formed with an sp hybridized orbital and two p-orbitals from each atom. The use of the p-orbitals forms a pi bond.

Carbon has the unique characteristic among all elements to form long chains of its own atoms, a property called catenation. This coupled with the strength of the carbon–carbon bond gives rise to an enormous number of molecular forms, many of which are important structural elements of life, so carbon compounds have their own field of study: organic chemistry.

Branching is also common in C–C skeletons. Different carbon atoms can be identified with respect to the number of carbon neighbours:

- Primary carbon atom: one carbon neighbour
- Secondary carbon atom: two carbon neighbours
- Tertiary carbon atom: three carbon neighbours
- Quaternary carbon atom: four carbon neighbours.

BONDING IN CARBON

That a chemical bond is essentially the sharing of electrons between two atoms. Since electrons are negatively charged and exert an attractive force on nuclei, they serve to hold the atoms together if they are located between two nuclei. How does this notion of a chemical bond relate to orbitals describing electron density? When two atoms approach each other, their atomic orbitals overlap. The overlapped atomic orbitals can add together to form molecular orbitals. This procedure is called the linear combination of atomic orbitals (LCAO). The area of greatest overlap between the original atomic orbitals represents the chemical bond that is formed between them. Since the sharing of electrons is the basis of the chemical bond, the molecular orbitals formed represent chemical bonds.

Let us take a simple example, hydrogen. We can imagine two 1s orbitals gradually coming closer and closer together until there is a good deal of overlap between them. At this point, the area in space of greatest electron density will be between the two nuclei, which themselves were at the center of the original atomic orbitals.

This electron density, now part of a new molecular orbital, represents the chemical bond. When, as below, the area of greatest overlap occurs directly between the two nuclei on internuclear axis or axis containing the nuclei of both atoms, the bond is called a sigma bond or ó bond.

There are two final points to consider when atomic orbitals combine. First, more than one atomic orbital from a single atom can be used to form new molecular orbitals. For example, a 2s orbital and a 2p orbital from one atom might add together and overlap with one or more orbitals from a second atom to form new molecular orbitals. Second, parts of orbitals can possess a sign (+ or -).

The s orbital has the same sign throughout, while in the p orbitals, one lobe is + and the other lobe is -. Signs do not matter with respect to electron density, but they must be taken into account when orbitals are added or subtracted. If two orbitals of the same sign are added, electron density will increase, while if two orbitals of opposite signs are added, the shared electron density will cancel out.

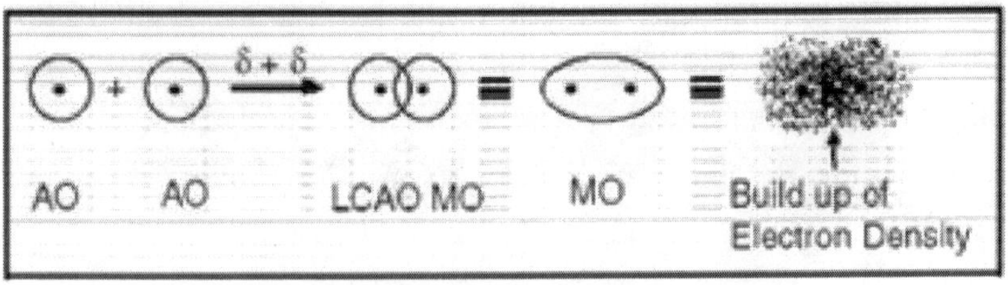

Fig. Two Hydrogen 1s Atomic Orbitals Overlap to form a Hydrogen Molecular Orbital

SP³ BONDS

Carbon has six electrons. Only two electrons can occupuy an s orbital at a time. The first two electrons in carbon occupy the 1s orbital. The next two occupy the higher-energy, but similarly shaped 2s orbital. The final two electrons occupy the 2p orbitals. In carbon, the electrons in the 1s orbital are too low in energy to form bonds. Thus, electrons used to form bonds must come from the 2s and 2p orbitals. Carbon very often makes four bonds, and when it does so, these bonds are arranged so that they are as far away from each other as possible. This arrangement is referred to as tetrahedral.

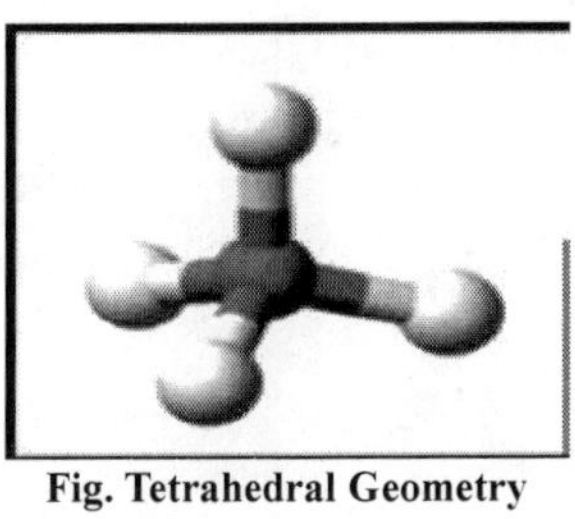

Fig. Tetrahedral Geometry

It is clear that the individual 2s and 2p orbitals cannot form bonds in this arrangement due to their geometry. The 2s orbital is completely symmetric, while the 2p orbitals are aligned along specific axes. None of these orbitals is well-equipped to form bonds in the tetrahedral geometry alone.

However, we have stated that a chemical bond doesn't have to be formed from individual atomic orbitals, but can be formed from a combination of several atomic orbitals from the same atom. For each bond that is made in the tetrahedral geometry, a part of the 2s and a part of each of the 2p orbitals will contribute, resulting in a tetrahedral arrangement. To achieve this geometry, both the 2s and all three of the 2p orbitals ($2p_x$, $2p_y$, and $2p_z$) must contribute.

The new bonds that are formed are called sp3 bonds, since one s and 3 p orbitals were used to form them.

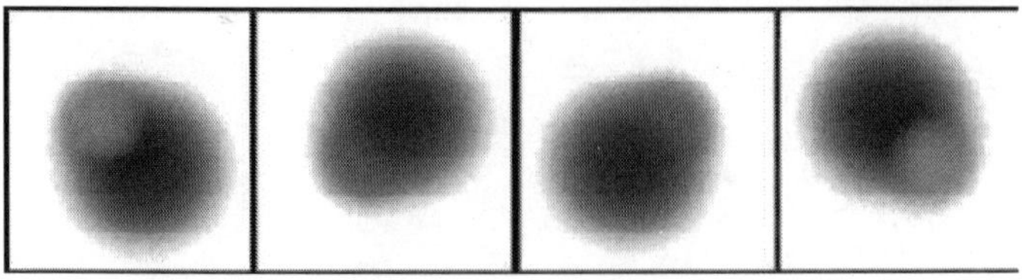

Fig. Orbital Combinations Used to make sp3 Bonds

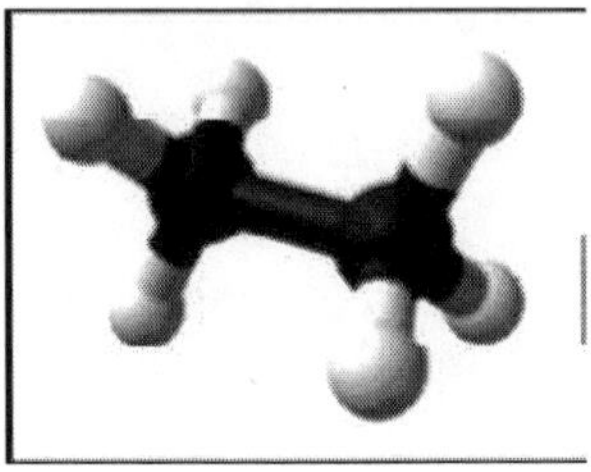

Fig. sp^3 Bond Between two Carbon Atoms

Note that when sp3 bonds are formed, they all point in separate directions as far away from each other as possible, leading to 109.5° angles between the bonds. This is the tetrahedral geometry as shown.

When carbon makes four sp^3 bonds to hydrogen, the resulting molecule is tetrahedral in shape. Carbon can also make a bond to another carbon atom. Two carbon atoms, both of which have sp^3 bonds oriented tetrahedrally, can bond to each other if each uses one of its sp^3 bonds.

SP² BONDS

Carbon sometimes makes just three bonds. In this case, not all of the 2p orbitals combine with the 2s orbital to form bonds. Instead, a combination of the 2s orbital and two of the 2p orbitals make three sp^2 bonds, while the other p orbital does not participate in this combination and can make a fourth bond on its own.

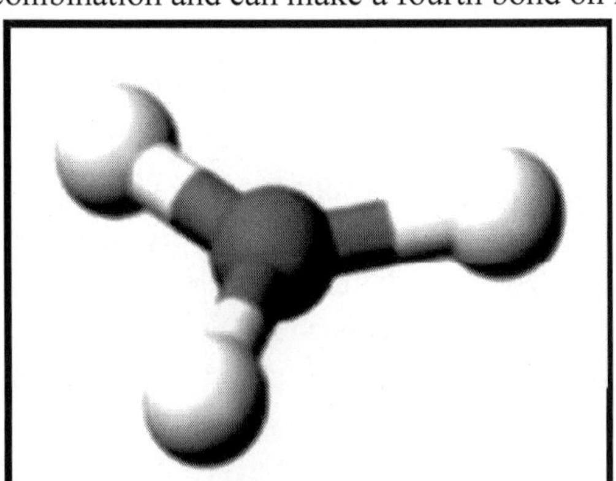

Fig. Trigonal Planar Geometry

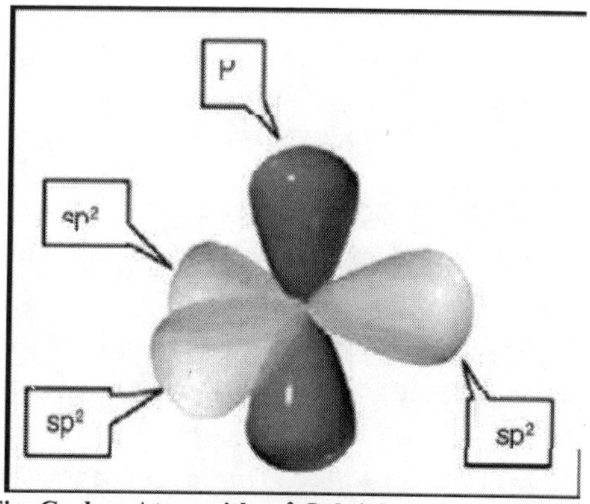

Fig. Carbon Atom with sp^2 Orbtials and a 2p Orbital

Like the sp3 bonds, the sp^2 bonds are oriented such that they are as far away from each other as possible. This geometry is called the trigonal planar geometry. It is called trigonal planar since each of the bonds points to one of the vertices of a triangle, but all three bonds are located in the same plane. What about the other 2p orbital, the one which did not add to make sp^2 bonds? This orbital still exists, and it is perpendicular to the plane in which the sp^2 bonds form. It too is able to form bonds, and it does so independently of the sp^2 bonds.

When two carbon atoms with sp^2 orbitals form a bond to each other using their sp^2 orbtials, a ó bond is formed between then. Morever, the extra p orbtials that exist above and below each carbon atom also overlap with each other. This orverlap between p orbitals leads to the formation of a second bond in addition to the ó bond formed between the sp^2 orbitals. This second bond, which does not occur directly between the nuclei on the internuclear axis, but rather above and below the internuclear axis, is called a pi bond or ð bond. When a ó bond and a ð bond form together between two atoms, a double bond is said to have formed.

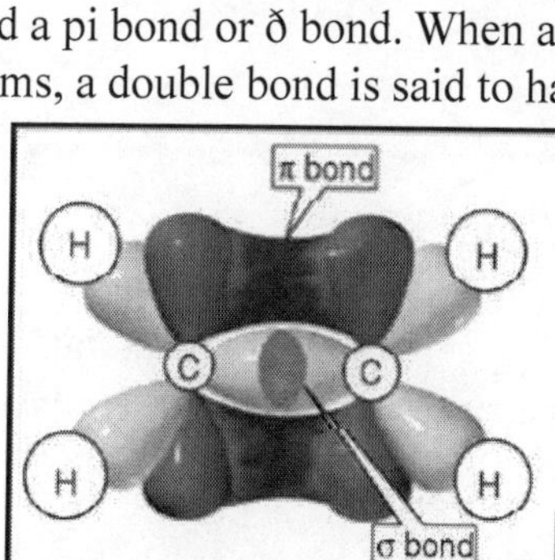

Fig. The Molecule Ethane is Formed From two Carbon Atoms.

As bond formed fromtwo sp2 orbitals and a Õ bond formed from two 2p orbitals together compose a double bond.

LEWIS DOT STRUCTURES

For the elements in the 2nd and 3rd periods, the number of valence electrons range from 1 to 8. Lewis dot structure for them are as indicated:

Li Be.B..C..N::O::F::Ne:

Using dots, Lewis made the valence electron visible. The stability of noble gases is now associated with the 8 valence electrons around it. The stability of 8 valence electrons led him to conclude that all elements strive to acquire 8 electrons in the valence shell, and the chemical reaction takes place due to elements trying to get 8 electrons. This is the octet rule. For the hydrogen and helium atoms, 2 electrons instead of 8 are required.

For example, the octet rule applies to the following molecules:

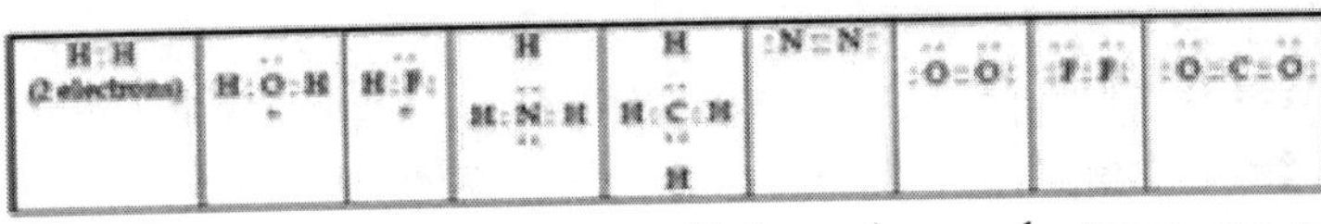

To draw a Lewis dot structure, all the valence electrons are represented. A good way is to draw a type of dot for the valence electrons of one atom different from types in another. To do this on the computer screen using only type fonts is difficult, but you should draw a few by hand on paper.

When a dash is used to represent a bond, it represents a pair of electrons. Thus, in the following representations, a dash represents two electrons, bonding or lone pairs.

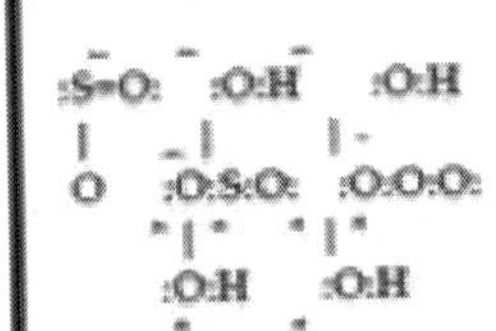

These structures satisfy the octet rule. Note the two ways of drawing the structures of H_2SO_4.

EXCEPTIONS TO THE OCTET RULE

Elements in the 3rd and higher periods may have more than 8 valence electrons. A possible explanation for this is to say that these atoms have d- type atomic orbitals to accommodate more than 8 electrons. In the following molecules, the number of valence electrons in the central atoms are as indicated:

Molecule	**SF_6**	**PC_{15}**	**IC_{13}**	**XeF_4**
No. of valenceelectronsforcentral atom	12	10	10	12

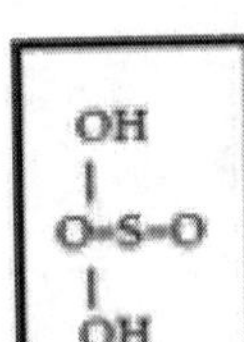

Draw the Lewis dot structures for the above molecules, and count the number of valence electrons for the central atoms. For H_2SO_4, the S atom has 12 electrons in the structure shown on your right. Each dash represent a chemical bond, which has two electrons. There is a total of 6 bonds around the S atom, and therefore 12 electrons. When B, Be, and some metals are the central atoms, they have less than 8 valence electrons. The following compounds do form, but the octet rule is not satisfied. These are electron defficient molecules.

Molecule	**$BeCl_2$**	**BF_3**	**BCl_3**	**$SnCl_2$**
No. of valenceelectronsforcentral atom	4	6	6	6

Another case of exception to the octet rule are molecules with odd number of valence electrons. For example:

No	**NO_2**	**ClO_2**	
No. of valenceelectronsforcentral atom	11	17	19

ISOELECTRONIC MOLECULES AND IONS

Counting the number of valence electrons often help us understand the formation of many molecules and ions. For example, all the following molecules have the 11 valence electrons:

NOCO-$O_2^+$$N_2^-$

The charged molecule do exist under special circumstance. The molecules of O_2 are paramagnetic, and thus, they have unpaired electrons. The first dot structure does not agree with this observed fact, but the second one does. However, the second one does not obey the octet rule.

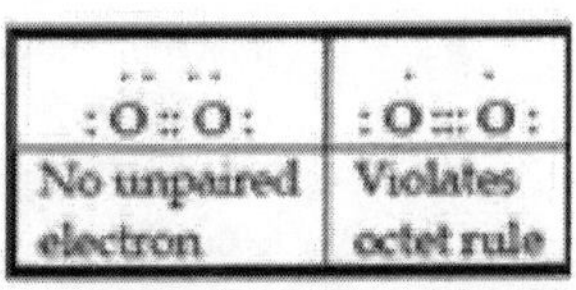

Later, you will learn that the molecular orbital (MO) theory provides a good explanation for the electronic configuration for O2.

RULES FOR ELECTRON DOT STRUCTURES AND BONDING STRUCTURES

The central atom follows the Octet Rule (usually) and in most cases the least electronegative non-metal is surrounded by the other atoms. Other atoms follow the Octet Rule whenever electrons are available. Check to see that every atom has the influence of 8 electrons and the total number of electrons is correct for that molecule.

Drawing bonding structures (called Lewis structures):

- Select a reasonable "skeleton" for the molecule or polyatomic ion
  - The LEAST electronegative element is usually the central element,

except that hydrogen never is example: S C S in compound CS_2

- Oxygen atoms do not bond to each other except in a few cases such as O_2 and O_3

- Calculate the total number of outer shell electrons available in all the atoms of the molecule or ion
- Draw a single bond to represent each pair of shared electrons in the skeleton
- Allowing 2 electrons for each shared pair, subtract the total number of electrons already used
- Distribute the remaining electrons in such a fashion as to give each element an octet, if possible
- For ions, be sure to add (for negative ions) or to subtract (for positive ions) the number of electrons indicated by the charge on the ion
- Remember that you can use double or triple bonds in order to give elements an octet, but only when necessary
- If there are any electrons "left over", place these additional unshared (lone) pairs of electrons into the skeleton to fill the octet of every group 1,2, 13, 14, 15, 16, 17 element (except hydrogen, which can only share 2 electrons). ❐

5.

Atoms and Atomic Spectroscopy

THE ATOMIC CONCEPT

Dalton first recognized and effectively utilized the stimulating force of the atomic concept and its ability to guide future physical research. To him we owe the use of the idea of atoms for understanding basic regularities of chemistry. Chemistry makes a sharp distinction between mixtures of different substances and chemical combinations. A mixture — as, for example, a solution of sugar in water or a mixture of nitrogen and oxygen — can include arbitrary (within certain limits) relative proportions of the constituents.

In a nitrogen-oxygen mixture the portion of oxygen can be varied continuously by addition or removal; correspondingly, the mixture exhibits properties somewhere between those of pure oxygen and of pure nitrogen, depending upon the relative concentration. Also, the mixture can be separated into its constituents through relatively superficial means. In a chemical combination entirely new properties appear, which can be quite dissimilar to those of the constituents. The proportions of the components of a chemical compound are quite stable and cannot be varied continuously.

Thus exactly 16 grams of oxygen always combine with 2.016 grams of hydrogen to form water. Another chemical substance, hydrogen peroxide, can be formed from 2.016 grams of hydrogen and just 32 grams of oxygen. It is striking that exactly twice as much oxygen is utilized in the second case as in the first; and on the basis of the atomic concept Dalton found a clear explanation for such facts, which occur similarly in all chemical combinations.

His explanation was as follows: The "molecule" of water, *i.e.*, the smallest possible particle of water, contains only one atom of oxygen; the molecule of hydrogen peroxide however contains two atoms of oxygen, while it contains just as many (two) atoms of hydrogen as the water molecule. Here we arrive at the distinction between atoms and molecules; we designate the smallest particles of chemical compounds as molecules, which for their part consist of atoms of the chemical elements. According to Dalton's ways of thinking a comprehensive study of proportions by weight in chemical compounds rendered the determination of the relative atomic weights of all the elements possible.

The "atomic" weight of oxygen is arbitrarily designated by the value 16, whereupon hydrogen acquires the atomic weight 1.008 and every other element, similarly, receives a definite atomic weight derived from chemical weight relationships. The regularities which the gases exhibit, which in the case of sufficiently small densities assume an "ideal" form, proved very helpful for the clarification of these ideas; for example, in the sense that a gas mass contracts to one-half its original volume when, with constant temperature, the pressure impressed upon it is doubled.

Criteria existed for the hypothesis that not only in compounds but also in a gaseous element such as nitrogen the particles moving at random through space are not necessarily identical with atoms; in nitrogen, for example, each "molecule" of the gas consists of two atoms. In the chemical combination of gases (with constant pressure and temperature) there is a simple relation between proportions by weight and volume relations (proportions by volume) between the chemical compound and the still uncombined constituents.

To fit these facts Avogadro introduced the explanation that equal volumes of all gases measured at constant pressure and temperature always contain the same number of molecules. Based on this "Avogadro's principle" a "molecular weight" can be determined for each gas independent of chemical weight relations.

The molecular weight of a chemically homogeneous gas is equal to the mass of 22.4 litres of this gas measured in grams at one atmosphere of pressure and a temperature of 0° C. Experience shows that values obtained in this way correspond with the chemically defined atomic weights. In many elements, *e.g.*, the metals, this molecular weight is exactly the same as the atomic weight; thus metallic vapours are "monatomic". In the aforementioned nitrogen, however, the molecular weight is exactly double the atomic weight. And in water it equals 18.016 ($2 \times 1.008 + 16$) in accordance with its above-described chemical composition.

STRUCTURE OF ATOM

Just as in the atom of hydrogen, in more composite atoms electrons can rotate around of a kern on allowed orbits only. The orbits in composite atoms cluster in system of shells. Each shell contains a particular number of orbits, an electron can be on any og them. A *K*-shell is the closest to a kern and contains two orbits. The second is a *L*-shell has eight orbits. The third is a *M*-shell has eight orbits, the fourth *N*-shell has 18 orbits etc.

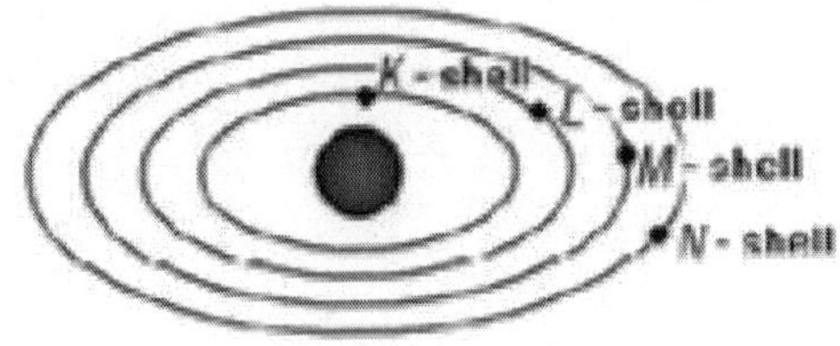

Parameters of Atom

When the electron travels from an orbit of greater radius to an orbit of smaller radius, the energy is oozed. But the transition from outer shells on interior is possible only when there are free orbits on the leter. The outer-shell electrons are less bound to a kern than interior ones: they are on greater distance from a kern, and the electrons, which are on inner shells, screen the nuclear charge. Therefore just a small amount of energy is sufficient for excitation of outer-shell electrons. The quantums of energy radiated at transition of such electrons in a non-excited state have a quantity of about several electron-volt. Greater energy is necessary to detach inner-shell electrons from the atom. The greater the nuclear charge the greater energy is needed. Therefore, when there is a transition of electrons from outer shells to a *K*-shell in heavy atoms, the emission of quantums with energy of hundreds and even thousand electron-volt is possible. Such quantums of high energy are called Roentgen rays.

X-ray Spectrums of Atoms

The X-rays results from losses by atom of an inner-shell electron: The transition of electrons is accompanied by an emission of quantums of high energy with a small wave length, which are termed as Roentgen rays. The X- rays emerges when atom losses inner-shell electron: The transition of electrons is accompanied by an emission of quantums of high energy with a small wave length called Roentgen rays. The atom can lose inner-shell electrons as a result of an electron bombardment with electrons speeded up to energy of several thousand electron-volt. The wave length of the X-rays is comparable to interatomic distance in solid bodies.

The examination of the X-ray diffraction on crystal lattices of solid bodies enables, knowing a wave length of Roentgen rays, to determine interatomic distances in them. The examination of Roentgen rays also allows to determine nuclear charge, because it depends on a prime relation of an electron-binding energy in atom, which is expressed through energy of the X-rays. This relation is called Moseley's law:

$$v = A\,(Z - B)$$

v and -frequency of a line in a X-ray charateristic spectrum of a device

A and B - constants Z -serial number of a device

PROPERTIES OF LIGHT

In one phenomena, which are bound to distribution of light, its undular properties (interference, diffraction) are shown. In other phenomena its quantum properties (photoeffect, radiation and uptake of a light, process of interaction of a light with substance) are shown. The light has simultaneously undular and quantum properties, which are differently shown in the different phenomena. However, when the wave length of a light is diminished, its undular properties are attenuated and the quantum properties are shown.

Compton's Effect

The experience of Arthur Holly Compton (1892-1962) on a dispersion of Roentgen rays by substances, which consist of atoms of light devices, have shown, that the dispelled Roentgen rays have a major wave length l¢, than impinging l. Thus the residual $\Delta\lambda = \lambda' - \lambda$ depends on properties of scattering substance and wave length of an impinging light:

$$\Delta\lambda = \lambda' - \lambda = 2\lambda_K \sin^2\frac{\theta}{2}$$

The quantity λ_K is a stationary value for all substances also is termed as Compton's wave length, $\lambda_K = 2{,}43 \times 10^{12}$ m. This phenomenon, which is termed as Compton's recoil, is impossible to explain by a wave theory of light. In quantum theory the magnification of wave length of radiation at a dispersion is explained by interaction of photons of impinging radiation with electrons of atoms and molecules.

The part of energy of an impinging photon, which is transmitted to electrons of scattering substance, depends on a scattering angle of photon.

$$At\theta = \pi,$$

that is provided that after a dispersion a photon will fly in the party, which is opposite to a tentative direction, the losses by a photon of energy will be maximal, that will give in maximal magnification of a wave length of radiation.

Light Pressure

Some properties, for example light pressure, can be surveyed simultaneously in the electromagnetic theory and in the quantum theory. According to the electromagnetic theory of light, the pressure p, which renders a luminous flux with power W, impinging normally on a simple surface, is equal:

$$p = \frac{W}{c}(1 + R)$$

For an ideal black body ($R = 0$): $p = \frac{W}{c}$

For an absolute reflecting bodyh ($R = 1$) $p = \frac{2W}{c}$

If N of quantums of a luminous flux with frequency n, which bears(carries) energy W = $N \cdot h$n, impinges on a simple surface for 1 second, then the blanket impulse at a reflectivity R will look like:

$$p = \frac{N.hv}{c}(1+R) = \frac{W}{c}(1+R)$$

Thus, electromagnetic theory and quantum theory give identical results. However it is essential different representations will be utillized in the electrom a gnetic and quantum theories of light. The concept of anelectromagnetic wave of application to a light guesses existence of a wave in particular field(area) of space, which is capable to contain even some lengths of waves of light.

The light quantums are considered as a particle, which are localized in very small volume of space. The representation about a light wave guesses a continuous distribution of energy in a spatial wave, the representation about quantums guesses localization of energy in a small part of space. Both these incompatible representations reflect unity of a wave-corpuscle nature of light, which is shown at distribution of light and its interaction with substance.

PROPERTIES OF SUBSTANCE

De Broglie's Theory

The French physicist Louis Victor duc de Broglie has applied wave- corpuscle treatment and to particles of substance: as the light detects corpuscular properties, also particles of substance should detect undular properties. De Broglie has utillized already wave-corpuscle properties, accepted for a light, and their quantum rule and has expressed a wave length through the performances of a particle. For a light the corpuscular point of view determines energy and impulse accordingly by expressions $E = mc^2$ and $p = mc$.

The wave length of a particle of substance is determined under the formula:

$$\lambda = \frac{h}{p} = \frac{h}{mv}$$

At adding in a rule of quantization of a moment of momentum of an electron the impulse of an electron, will turn out a relation $n\lambda = 2\pi r$. From it follows, that on length of a stationary orbit of an electron the integer of lengths of waves should be stacked. Thus, the requirement of existence of allowed orbits, which was injected by Bohr frequency relation, acquires legible physical sense: the allowed orbits are orbits supposing formation on them of standing waves of an electron. For a rectilinear motion of particles de Broglie has offered expression, which is made by analogy with the equation for a spread flat light wave in the complex shape:

$$\psi = \psi_o ei\left(wt - \frac{x}{\lambda}\right)$$

De Broglie's Waves

The specified waves have received a title of associated waves. Thus, the associated waves are traveling waves for freely propellent electrons or standing waves for electrons, which are bound in atoms. De Broglie's waves are not waves of propellent substance, they have not analog in classical physics. De Broglie's guess about existence of undular properties of particles carries universal character: undular properties should have the electron, positive proton, neutron, atom, molecule, any moving object.

However objects, which have major masses and move with usual velocities, will have a so small wave length To in comparison with the sizes of objects, that the phenomena of an interference and diffraction for them can completely be neglected. The undular properties of a light are clearly shown in cases, when the wave length has compared to the sizes of bodies, with which the light interreacts. The lengths of waves of electrons in usual requirements have the order of the nuclear sizes, hence, the effects are characteristic for them which are observed usually for Roentgen rays. For such small lengths of waves it is possible to observe a diffraction on nuclear crystal lattices.

EXAMINATION OF PROPERTIES

The existence of undular properties of substance was revealed at observation of patterns of a diffraction of beams of particles. Legible patterns of an interference and the diffractions give bundles of electrons, which have almost identical impulses. These patterns are very similar to optical patterns of an interference and diffraction. At change of quantity of a momentum of electrons and measuring of a wave length, it is possible to receive a relation between these quantities. It has appeared equal to a known relation;

$$\lambda = \frac{h}{p}$$

This de Broglie's relation is valid for any kind of particles. The undular properties of particles were detected at observation of a pattern of the diffraction of the bundle of electrons on chips. At measuring a scattering angle it is possible to spot a wave length of electrons.

Thus, at known quantity of potential, which was utilised for acceleration of a bundle of electrons, their velocity is possible to calculate, and then to receive independent value of a wave length of an electron. The value of a wave length of an electron, which was spotted in such a way, completely coincides with quantity of a wave length, which is found from a diffraction pattern. The undular properties of electrons also were detected in experience with a cathode rays at major velocity, which was passed through a thin metal foil on a photographic plate. After development of a photoplate on it the diffraction pattern from rings of different intensity was visible. However this diffraction pattern was stipulated not by Roentgen rays, which could arise at a concussion of high-velocity electrons with metal. It confirms presence of a stream of charged particles.

MECHANICS

Quantum Mechanics is the study of matter and energy on the subatomic scale. It was came into being because classical physics could not explain certain experimental results, such as why the photoelectric effect occured only when the light shone is above a certain frequency. Quantum Mechanics and Relativity became the foundation of modern physics.

Main Points

Principle of Quantatization

The fundame ntal principle of Quantum Mechanics, and its namesake, is that energy came in discrete particles called "quanta". This idea, the Quantatization of Energy, was put forth by Planckinhis Quantum Hypothesis(1900), an explanation of blackbody radiation. It was a departure from Classical Physics, which asserted that energy could be infinately small.

Photoelectric Effect

This quantatization of energy approach was used by Einstein in his 1905 explanation of the photoelectric effect, the emission of electrons by a metal when light is shone on it. No electrons are emitted when the light is below a threshold frequency, no matter how intense the light. Einstein postulated that light, while exibiting wave properties as shone by the Double-slit Experiment, also came in the from of particles called "photons", discrete packets of energy. The energy of a photon is proportional to the frequency of the light($E=hv$); thus, light with higher frequency had more energy and could excite the electrons, while light whose frequency were below the threshold could not, no matter how many photons there were.

Bohr's Model and Spectral Lines

Bohr also used the Quantatization of energy in his 1913 model of the atom.

He stated that electrons revolved around the nucleus much like the planets revolve around the sun. They could be found in predefined, stationary orbits, radiating no energy as they revolve. Only when they gain a quatum of energy equivalent to the difference between the energy of the orbits do they become excited and enter a higher energy orbit. This is consistent with Einstein's explanation of the photoelectric effect, and explains why spectral lines only occur at certain places.

Wave-Particle Duality of Matter

In 1923 De Broglie expanded the Principle of Wave-particle Duality of Light put forth by Einstein to include all matter, called Principle of Wave- particle Duality of Matter. He asserted that all matter, not just light, exibit both wave-like and particle-like behaviour,

Quantum Matrix and Wave Mechanics

By now there was enough theories to put together a quantum picture of the atom. However, it was very difficult to sort through all the math. Heisenherg and Shroedinger simultaneously and independently worked to put forth a complete quantum mechanics. Heisenberg employed use of matrices, and his model was called the matrix mechanics. His approach emphasized the quantum — discrete properties while Schroe dinger concentrated on the wave properties. He tried to find a function that would encompass all informatin about a particle or system at any given time. This is called the wave function. While debated ensued about which was correct, Schoedinger proved shortly after that the matrix mechanics and the wave function were mathematically equivalent.

Uncertainty Principle

Heisenberg then came up with the Uncertainty Principle, which stated that the position and velocity(momentum, which is mass*velocity) cannot be known precisely simultaneouely. The more accurately one is known, the less accurately the other can be measured. The product of the uncertainty in position, and the uncertainty in momentum, is always greater than or equal to h/(2pi). While this is neglegible on the macroscopic level, subatomically, it causes great uncertainty. As a result, the world exists in statistical probabilities. For example, a electron cannot be found precisely at one location. At any given chance, it has a certain probability of being found anywhere in the universe, with higher probabilities in regions around the atomic nucleus and less likelihood elsewhere. This can be calculated with the wave function.

Quantum Atomic Model

The quantum model of the atom calls for arbitrarily defined orbitals. An orbital is a region around the nucleus where an electron has 90 per cent chance of being found. Electrons have spin, a term used for the intrinsic angular momenta subatomic particles have. They have either a spin of +1/2 or -1/2. Two electrons in the same orbital of the same energy level must have opposite spin (Pauli Exclsion Principle).

Schroedinger's Cat

Since the wave function describes the world in terms of statistical probabilities and not certainties, a cat placed in an isolated system with equal chances of being found alive and dead at anytime *is* equally dead and alive at the same time. When an observer is introduced to the isolated system, the wave function collapses and the cat is either dead or alive.

THE NEUTRAL ATOMS

The ground-state electron configurations of elements heavier than neon are shortened in the table by using rare-gas element symbols in brackets to represent the corresponding electrons. The ground levels of all neutral atoms have reasonably me an in g ful LS -coupling names, the corres ponding eigenvector percentages lying in the range from ~55 per cent to 100 per cent. These names are listed in the table, except for Pa, U, and Np; the lowest few ground-configuration levels of these atoms compri se better $5f^N(L_1S_1J_1)$, $6d_j7s^2$ $(J_1 j)$ terms than LS-coupling terms. The relatively large spin-orbit interaction of the $6p$ electrons produces jj-coupling structures for the $(6p^2{}_{1/2})0$, $(6p^2{}_{1/2}6p_{3/2})o_{3/2,}$ and $(6p2_{1/2}6p2_{3/2})_2$ ground levels of the $6p^2$, $6p^3$, and $6p^4$ configurations of neutral Pb, Bi, and Po, respectively. As noted in the chapter *jj* Coupling of Equivalent Electrons, the *jj*-coupling names are more appropriate for these atoms than the alternative LS-coupling designations in the table.

The ionization energies in the table are from recent compilations. The uncertainties are mainly in the range from less than one to several units in the last decimal place, but a few of the values may be in error by 20 or more units in the final place; *i.e.*, the error of some of the two place values could be greater than 0.2 eV. Although no more than four decimal places are given here, values for both the neutral and singly-ionized atoms are given to their full accuracies in.

ZEEMAN EFFECT

The Zeeman effect for "weak" magnetic fields (the anomalous Zeeman effect) is of special interest because of the importance of Zeeman data in the analysis and theoretical interpretation of complex spectra. In a weak field, the J value remains a good quantum number although in general a level is split into magnetic sublevels. The g value of such a level may be defined by the expression for the energy shift of its magnetic sublevel having magnetic quantum number M, which has one of the $2J + 1$ values, $-J, -J+1, ..., J$:

$$\Delta E = gM\mu_B B$$

The magnetic flux density is B, and mB is the Bohr magneton . $\left(\mu_B = e\hbar / 2m_e\right)$. The wavenumber shift $\Delta\sigma$ corresponding to this energy shift is,

$$\Delta\sigma = gM\left(0.46686\ B\ cm^{-1}\right),$$

with B representing the numerical value of the magnetic flux density in teslas. The quantity in parentheses, the Lorentz unit, is of the order of 1 or 2 cm^{-1} for typical flux densities used to obtain Zeeman-effect data with classical spectroscopic methods. Accurate data can be obtained with much smaller fields, of course, by using high er-res olution techniques such as laser spectroscopy. Most of the g values now available for atomic energy levels were derived by application of the above formula (for each of the two combining levels) to measurements of optical Zeeman patterns. A single transverse-Zeeman-effect pattern (two polarizations, resolved components, and sufficiently complete) can yield the βJ value and the g value for each of the two levels involved.

Neglecting a number of higher-order effects, we can evaluate the g value of a level βJ belonging to a pure LS-coupling term using the formula

$$g_{\beta SLJ} = 1 + (g_{e-1})\frac{J(J+1) - L(L+1) + S(S+1)}{2J(J+1)}$$

The independence of this expression from any other quantum numbers (represented by β) such as the configuration, etc., is important. The expression is derived from vector coupling formulas by assuming a g value of unity for a pure orbital angular momentum and writing the g value for a pure electron spin as g_e. A value of 2 for g_e yields the Landé formula. If the anomalous magnetic moment of the electron is taken into account, the value of g_e is 2.0023193. "Schwinger" g_e values obtained with this more accurate value for g_e are given for levels of SL terms in Ref.. The usefulness of g_{SLJ} values is enhanced by their relation to the g values in intermediate coupling. In the notation used in Equation for the wave function of a level $\hat{a}J$ in intermediate coupling, the corresponding g value is given by

$$g_{\beta J} = \sum_{\gamma^{SL}} gSLJ \backslash c\left(\gamma^{SLJ}\right)^2$$

where the summation is over the same set of quantum numbers as for the wave function. The $g_{\beta J}$ value is thus a weighted average of the Landé g_{SLJ} values, the weighting factors being just the corresponding component percentages.

THE STRUCTURE OF THE ATOM AND ELECTRON

The discovery of the electron in 1897 by J. J. Thomson suggested that there was an internal structure to the "indivisible" building blocks of matter known as atoms. This raised an obvious question: How many electrons does an atom contain? By studying the scattering of light, x-rays, and -particles, Thomson concluded that the number of electrons in an atom was between 0.2 and 2 times the weight of the atom.

In 1911, Rutherford concluded that the scattering of -particles by extremely thin pieces of metal foil could be explained by assuming that all of the positive charge and most of the mass of the atom were concentrated in an infinitesimally small fraction of the total volume of the atom, for which he proposed the name nucleus. Rutherford's data also suggested that the nucleus of a gold atom carries a positive charge that is about 80 times the charge on an electron.

The discovery of the neutron in 1932 explained the discrepancy between the charge on the nucleus and the mass of an atom. A neutral gold atom that has a mass of 197 amu consists of a nucleus that contains 79 protons and 118 neutrons surrounded by 79 electrons. By convention, this information is specified by the following symbol, which describes the only naturally occurring isotope of gold.

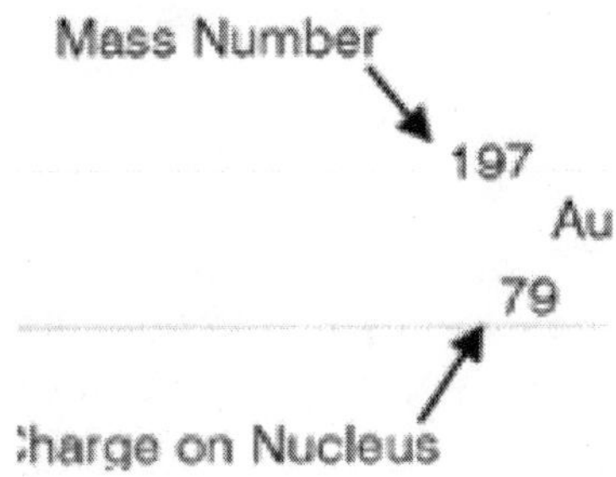

This convention can also be applied to subatomic particles. The only difference is the use of lowercase letters to identify the particle. Because anyone with access to a periodic table can find the atomic number of an element, a shorthand notation is often used that reports only the mass number of the atom and the symbol of the element. The shorthand notation for the naturally occurring isotope of gold is ^{197}Au.

RADIOACTIVE DECAY

Early studies of radioactivity indicated that three different kinds of radiation were emitted, symbolized by the first three letters of the Greek alphabet, β and. With time, it became apparent that this classification scheme was much too simple.

The emission of a negatively charged β particle, for example, is only one example of a family of radioactive transformations known as β-decay. A fourth category, known as spontaneous fission, also had to be added to describe the process by which certain radioactive nuclides decompose into fragments of different weight.

ALPHA DECAY

Alpha decay is usually restricted to the heavier elements in the periodic table. (Only a handful of nuclides with atomic numbers less than 83 emit an - particle.) The product of -decay is easy to predict if we assume that both mass and charge are conserved in nuclear reactions. Alpha decay of the ^{238}U "parent" nuclide, for example, produces ^{234}Th as the "daughter" nuclide.

$$^{238}_{92}U \rightarrow \, ^{234}_{90}Th + \, ^{4}_{2}He \ (\alpha\text{-decay})$$

The sum of the mass numbers of the products (234 + 4) is equal t nass number of the parent nuclide (238), and the sum of the charges on the products (90 + 2) is equal to the charge on the parent nuclide.

BETA DECAY

There are three different modes of beta decay:

- Electron (β^-) emission
- Electron capture
- Positron (β^+) emission

Electron (β^-) emission is literally the process in which an electron is ejected or emitted from the nucleus. When this happens, the charge on the nucleus increases by one. Electron (β^-) emitters are found throughout the periodic table, from the lightest elements (^{3}H) to the heaviest (^{255}Es).

The product of β^-emission can be predicted by assuming that both mass number and charge are conserved in nuclear reactions. If ^{40}K is a β^--emitter, for example, the product of this reaction must be ^{40}Ca.

$$ {}^{40}_{19}K \rightarrow {}^{40}_{20}Ca + {}^{0}_{-1}e \quad \left[electron(\beta^-)emission\right] $$

Once again the sum of the mass numbers of the products is equal to the mass number of the parent nuclide and the sum of the charge on the products is equal to the charge on the parent nuclide. Nuclei can also decay by capturing one of the electrons that surround the nucleus. Electron capture leads to a decrease of one in the charge on the nucleus.

The energy given off in this reaction is carried by an x-ray photon, which is represented by the symbol hv, where h is Planck's constant and v is the frequency of the x-ray. The product of this reaction can be predicted, once again, by assuming that mass and charge are conserved.

$$ {}^{40}_{19}K + {}^{0}_{20}e \rightarrow {}^{40}_{-1}Ar + hv \ (electron\ capture) $$

The electron captured by the nucleus in this reaction is usually a 1s electron because electrons in this orbital are the closest to the nucleus. A third form of beta decay is called positron (β^+) emission. The positron is the antimatter equivalent of an electron. It has the same mass as an electron, but the opposite charge. Positron

(β^+) decay produces a daughter nuclide with one less positive charge on the nucleus than the parent.

$$^{40}_{19}K \rightarrow \, ^{40}_{18}Ar + \, ^{0}_{+1}e \quad [positron(\beta^+)emission]$$

Positrons have a very short life-time. They rapidly lose their kinetic energy as they pass through matter. As soon as they come to rest, they combine with an electron to form two g-ray photons in a matter-antimatter annihilation reaction.

$$^{0}_{+1}e + \, ^{0}_{-1}e \rightarrow 2\gamma$$

Thus, although it is theoretically possible to observe a fourth mode of beta decay corresponding to the capture of a positron, this reaction does not occur in nature. Note that in all three forms of b -decay for the 40K nuclide the mass number of the parent and daughter nuclides are the same for electron emission, electron capture, and position emission. All three forms of b -decay therefore interconvert isobars.

GAMMA EMISSION

The daughter nuclides produced by -decay or β-decay are often obtained in an excited state. The excess energy associated with this excited state is released when the nucleus emits a photon in the g-ray portion of the electromagnetic spectrum.

Most of the time, the g-ray is emitted within 10^{-12} seconds after the -particle or β-particle. In some cases, gamma decay is delayed, and a short-lived, or metastable, nuclide is formed, which is identified by a small letter m written after the mass number. ^{60m}Co, for example, is produced by the electron emission of ^{60}Fe.

$$^{60}_{26}Fe \rightarrow \, ^{60m}_{27}Co + \, ^{0}_{-1}e$$

The metastable ^{60m}Co nuclide has a half-life of 10.5 minutes. Since electromagnetic radiation car ries neither charge nor mass, the product of g- ray emission by ^{60m}Co is 60Co.

$$^{60m}_{27}Co \rightarrow \, ^{60}_{27}Co + \gamma \quad (\gamma\text{-ray emission})$$

SPONTANEOUS FISSION

Nuclides with atomic numbers of 90 or more undergo a form of radioactive decay known as spontaneous fission in which the parent nucleus splits into a pair of smaller nuclei. The reaction is usually accompanied by the ejection of one or more neutrons.

$$^{252}_{98}Cf \rightarrow {}^{140}_{54}Xe + {}^{108}_{44}Ru + 4\,{}^{1}_{0}n$$

For all but the very heaviest isotopes, spontaneous fission is a very slow reaction. Spontaneous fission of ^{238}U, for example, is almost two million times slower than the rate at which this nuclide undergoes α -decay.

THE HYDROGEN ATOM

The Schrödinger equation in 1-D is extended to 3-D in much the same way as the energy method in 1-D is extended to 2-D to deal with central-force motion.

The kinetic-energy term associated with the angular variable(s) is written as a function of r and the angular momentum l, and is treated as a form of potential energy. The resulting equation looks (somewhat in the case of the Schrödinger equation) like a 1-D problem with an "effective" potential energy.

The Bohr model of the hydrogen atom uses the classical techniques of central-force motion along with quantized angular momentum, and finds the energy levels successfully. It is a convenient starting point here because it provides the quantities we need to scale the Schrödinger equation into dimensionless form.

The three equations needed for the model are Newton's law:

$$\frac{mV^2}{r} = \frac{e^2}{4\pi\varepsilon_o r^2}$$

Energy Conservation:

$$\frac{mV^2}{2} = \frac{e^2}{4\pi\varepsilon_o r^2} = E$$

Quantization of the angular momentum:

$$l = mVr = n\hbar$$

where n has integral values 1, 2, 3.. The three equations can be solved for the radii and energies of the orbits:

$$r = n^2 \frac{4\pi\varepsilon_o \hbar^2}{me^2} = n^2 a_o$$

$$E = -\frac{me^4}{2(4\pi\varepsilon_o \hbar)^2}\frac{1}{n^2}$$

$$= -\frac{mc^2}{2}\left(\frac{e^2}{4\pi\varepsilon_o \hbar c}\right)^2 \frac{1}{n^2} = -\frac{mc^2\alpha^2}{2}\frac{1}{n^2} = -\frac{13.6eV}{n^2}$$

The dimensionless group of constants represented by a is known as the fine-structure constant and is approximately equal to 1/137. We will use the first orbit radius (approximately 0.053 nm) and energy as the length and energy units in the dimensionless Schrödinger equation.

The Schrödinger equation looks fairly complicated when written in spherical polar coordinates. The solution is written as the product of a function R of r and a function Y (known as a spherical harmonic) of the angular variables. There are elegant ways of dealing with spherical harmonics in the quantum mechanical theory of angular momentum, and there are even more elegant ways in group theory where they are basis functions for the rotation group in 3-D. (There is nothing "quantum mechanical" about the spherical harmonics.) We will say nothing more about the angular variables here beyond noting that they contribute the angular momentum term to the radial Schrödinger equation:

$$-\frac{\hbar^2}{2m}\frac{d^2}{dr^2}(rR) + \frac{l(l+1)\hbar^2}{2mr^2}(rR) - \frac{e^2}{4\pi\varepsilon_o r}(rR) = E(rR)$$

The angular momentum quantum number l has values 0, 1, 2.., and note the product $l(l + 1)$ in the angular momentum term rather than the *n2* of the Bohr model. It is not necessary to write the equation in terms of the product rR, but it is convenient. (The product has some physical significance: when it is squared it is proportional to the radial probability density, the probability density for a thin spherical shell at radius r.) In dimensionless form the equation becomes:

$$-\frac{d^2}{dr^2}(rR) + \frac{l(l+1)}{r^2}(rR) - \frac{2}{r}(rR) = E(rR)$$

This equation is not suitable for numerical solution because, even with one power of r cancelled, the angular momentum term goes to infinity at the origin. What this tells us is that near the origin,

$$R = r^l u$$

where u is non-zero at the origin. When we make this substitution,

$$\frac{d^2u}{dr^2} = -Eu - 2\frac{u + (l+1)du/dr}{r}$$

This is the equation we solve numerically using the Feynman algorithm. (The term in $1/r$ dictates the value of du/dr at the origin.)

The programme operates in two modes:

- It finds the energy levels for which the wavefunction tends asymptotically to zero at large *r*, and
- It plots the radial probability density.

The usual *s, p, d, f* notation is used for $l = 0, 1, 2, 3$. The horizontal axis is divided into four divisions, each equal to the radius of the *n*th Bohr orbit. Suitable vertical scale factors are input along with the *n* and *l* values for the selected state.

The Motivation for Quantum Mechanics

Physicists at the end of the nineteenth century believed that most of the fundamental physical laws had been worked out. They expected only minor refinements to get "an extra decimal place" of accuracy. As it turns out, the field of physics was transformed profoundly in the early twentieth century by Einstein's discovery of relativity and by the development of quantum mechanics. While relativity has had fairly little impact on chemistry, all of theoretical chemistry is founded upon quantum mechanics. The development of quantum mechanics was initially motivated by two observations which demonstrated the inadeqacy of classical physics. These are the "ultraviolet catastrophe" and the photoelectric effect.

IMPORTANCE OF RELATIVISTIC EFFECTS IN ATOMS

The importance of "relativistic effects" in properly describing the electronic structure of molecules containing heavy elements is often stressed in the literature. A logical definition of these effects might be: "The differences in molecular property predictions made by the relativistic Hamiltonian and the non-relativistic Hamiltonian". However, this definition is impractical for several reasons. The first problem with this definition is that it requires the existence of an exact relativistic Hamiltonian. This may be resolved by choosing a single, well-accepted approximate Hamiltonian such as the Dirac-Coulomb or Dirac-Breit Hamiltonian as *the* relativistic Hamiltonian. Another problem stems from the extensive discussion of "correlation effects" in the quantum chemical literature.

This term is used to describe the difference between properties predicted by the Hartree-Fock wavefunctions and the properties predicted by higher level, c orrelated w a vefunctions. In order to study relativis tic eff ects independently of correlation effects, it is useful to define relativistic effects as the differences between the predictions of the Hartree-Fock (HF) and Dirac- Hartree-Fock (DHF) methods. In this way, both relativistic effects and correlation effects may be viewed as corrections to the non-relativistic HF model.

It is important to note, however that the two types of effects should not be regarded as separate, additive effects, and that significant interplay between the two should be expected in cases where relativistic and correlation effects are significant.

Unfortunately, there are other approximations inherent in the HF and DHF methods structure methods which stand in the way of isolating those differences which are purely due to the introduction of relativity. The properties predicted by a particular molecular wavefunction are often highly dependent on the quality of the finite basis set in which the wavefunction is expanded. This is especially true in the case of the DHF method, where the reliability of the predictions are not only dependent upon the size and flexibility of the basis set, but also upon the proper balance of large and small component basis sets. The extraordinarily different basis set requirements of the two methods make it difficult to determine what would constitute an equal treatment of the two.

For atomic systems, however, the issue of basis set effects may be avoided by solving the one dimensional DHF and HF radial equations numerically. This makes atomic systems a good starting point for the investigation of isolated relativistic effects. Furthermore, characterization of the differences in such properties as orbital shape and extent, and orbital energies for isolated atomic systems can help to identify periodic trends in relativistic effects without the perturbation of a molecular environment. Models established in the high symmetry of the central force field of the atom may then lend structure to descriptions of the valence orbital effects in molecules, where significant orbital mixing may be encouraged by the lower symmetry molecular environment.

When molecular systems are too large to be treated with the relatively rigourous, DHF-based four-component methods, more approximate two- component methods such as relativistic perturbation theory are utilized in an attempt to include the most important relativistic effects. In these cases, it is essential to have an estimate of the magnitude of the difference between the relativistic and non-relativistic pictures in order to evaluate the validity of perturbative schemes.

DIRECT, INDIRECT AND SPIN-ORBIT EFFECTS

A crude estimate of the importance of relativity in characterizing the atomic orbital energies and spatial extent may be achieved by invoking the Bohr model of the hydrogenic atom of charge Z. In this classical picture, an electron in an orbit designated by principal quantum number n would travel at a radius r_n and a velocity v_n with a total energy E_n given (in a.u.) by

$$\begin{aligned} r_n &= \frac{n^2}{mZ} \\ v_n &= Z/n \\ E_n &= -\frac{mZ^2}{2n^2} \end{aligned}$$

where m has been retained intentionally. Since the electron is participating in accelerated motion, the equations of special relativity cannot apply rigourously. However, it is possible to achieve an estimate of the changes which a relativistic Hamiltonian would introduce for this system by replacing m with the velocity-dependent mass

$$m = \frac{m_o}{\sqrt{1-v^2/c^2}} = \frac{m_o}{\sqrt{1-Z^2/n^2c^2}}$$

The manifestation of this behaviour is typically classified as a direct relativistic effect. Direct effects are most obvious for the *s* and *p* type orbitals since these orbitals tend to have a greater density near the nucleus, and hence experience a greater fraction of the full nuclear charge.

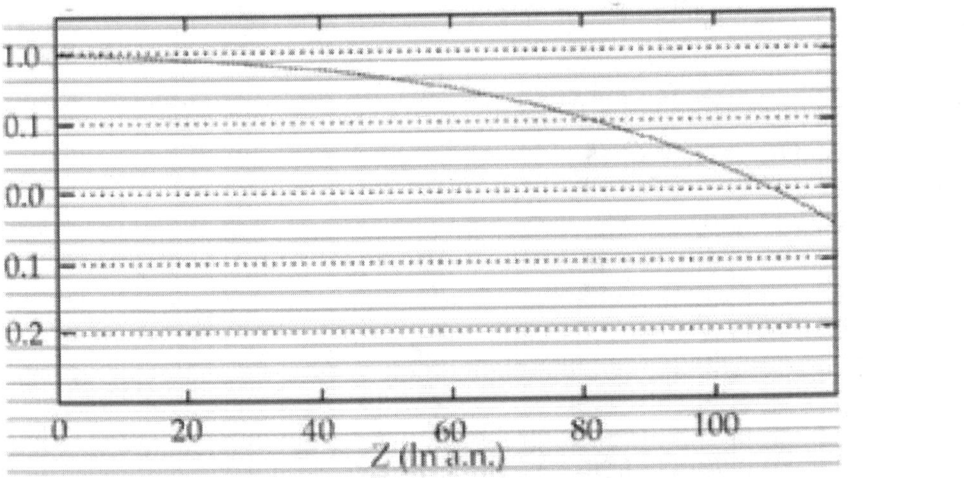

Fig. Ratio of Relativistic Bohr Orbit to Non-Relativistic Bohr Orbit

The orbital shrinking experienced by *s* and *p* orbitals will, in turn, have an affect on the higher angular momentum *d* and *f* orbitals. The greater density of the *s* and *p* orbitals in the nuclear vicinity effectively crowds out the *d* and *f* orbitals through orbital orthogonality and shielding effects. The resultant expansion of the *d* and *f* orbitals and the associated shift of the orbital energies is generally classified as an indirect relativistic effect. In atomic systems where this orbital shrinking and expansion is highly pronounced, non-relativistic methods will obviously be inadequate.

Another litmus test of the necessity of a relativistic treatment is the contribution of the small component to the total electron density predicted by the DHF method for a particular atomic orbital. The pattern of electron density and or though on a lity relation ships associated with a four - component wave function with appreciable large and small components are difficult to reproduce with a two-component wavefunction. The small component of the wavefunction tends to be largest very close to the nucleus and therefore plays a substantial role in the screening of nuclear charge for the higher energy orbitals. The contribution of the small component is usually greatest for orbitals with low principal quantum numbers. This is illustrated by Figure which depicts the large and small components of the radial wavefunction for the **s**- like orbitals (j = 1/2, k = – 1, l = 0) of radon with principal quantum numbers n = 1, 2, 3.

Similarly, the small component to the 1s orbital of increasingly lighter elements tends to be smaller, and the profile of the small components tends to more closely match that of the large component. Therefore, the renormalized large component should give a closer estimate of the radial distribution of the electron density for the lighter elements. This may be observed the plots of the large and small components of the 1s-like DHF orbitals of Rn, Xe, and Kr depicted.

Spin-orbit (S – O) coupling is one of the most manifest of relativistic effects in chemistry. A simple picture of S – O coupling may be obtained by returning to the Bohr atom and employing the classical expression for the interaction of the electron's spin and the magnetic field which results from its orbital motion. An electron orbiting a nucleus may be said to experience the classical magnetic field

$$B = \left(E \times v / c\right)$$

where E for a central potential $V(r) = \frac{z}{r}$ is given by,

$$E = \frac{1}{r}\frac{\partial V}{\partial r} r$$

The interaction energy, E_{s-o}, of the intrinsic magnetic moment of the electron with this magnetic field is given by,

$$\mu B = \left[-\left(\frac{e\hbar}{2mc}\right)\sigma\right].E \times p$$

$$= \frac{e\hbar}{2mc}\sigma.\left[\left(\frac{1}{r}\frac{\partial V}{\partial r} r\right) \times p\right] = \frac{e}{mc}\frac{1}{r}\frac{\partial V}{\partial r} l.s$$

where the definition $s = \frac{\hbar}{2}\sigma$ has been employed. This classical interaction term turns out to be too large by a factor of 2. The reason for this is directly related to the fact that our picture has assumed that the central nucleus is stationary. This classical picture provides a good model for the sorts of interactions which will give rise to the atomic fine structure. The Dirac-Coulomb Hamiltonian accounts for this sort of spin-orbit coupling in multi-electron atoms, but the interaction of the spin of an electron with the orbital motion of another electron and the interaction of one electron spin with that of another are not treated by the DC Hamiltonian. In order to take these effects into account, the Breit interaction term must be added. Fortunately, such interaction terms are typically very small, and do not have a large effect on atomic electronic structure, but they must be included for even qualitative predictions of the atomic spectra of the heaviest atoms.

RELATIONS OF MOLECULAR WEIGHTS

These ideas are further verified by simple relations of molecular weights such as, "diffusion velocities", or by the characteristic difference between the specific heats of "monatomic" and "polyatomic" gases. None of these, it must be emphasized, are proofs for the correctness of the atomic concept; but they are proofs for its usefulness. The atomic idea describes very plainly a large number of important and comprehensive regularities, thus fadlitating greatly our progress with these phenomena.

Crystallography furnishes evidence of further regularities which are certainly clearly explained by the atomic idea, though they too do not lead to proofs for the reality of the atom. In the crystalline state of matter the atomic concept suggests a regular grouping of the atoms or molecules, alongside of each other and arranged in layers. This view induces important results; on this basis the variety of possible forms of crystals are shown mathematically to be limited very considerably. The completion of this mathematical proposition yields the wonderful result that according to the atomic concept there should be no more nor less than 32 different "crystal classes" which are defined by different properties of symmetry.

Actually this confirms precisely the experience of mineralogists; there really occur in nature all the crystal forms (crystal symmetries) which are compatible with the atom idea, but not one single one which contradicts it. The so-called "law of rational indices" discovered by the mineralogists formed an important adjunct to this original consideration, which is related only to the symmetry properties of different crystal forms. If we imagine the crystal as built up of a regular stratification of layers of atoms, then its outer boundary surfaces could never lie so that they partially cut through the atoms.

The limits for the possible positions of the boundary surfaces given by this view are exactly equivalent to the requirements which the law of rational indices places on the position of the crystal surfaces. Such experiences must necessarily encourage the more and more energetic pursuit of the atomistic notion — which intrinsically was so clear — even though tangible proof for the reality of atoms was still lacking. Actually the theoretical development of the atomic concept was furthered extensively by the most able physicists. We have already mentioned ideal gases and indicated how the gaseous state of aggregation should be represented on the basis of the atomic theory; the molecules of a gas are completely separate from each other; each one moves along with high velocity until it meets another one, and then, in elastic rebound, their direction of motion and speed change.

The continuous impact of countless molecules on the walls of the enclosing vessel produces the pressure of the gas, perceptible to us with our rough tools. These ideas were elaborated thoroughly (Clausius, Maxwell) and our understanding of the gaseous state of aggregation was developed very clearly and completely. Whereby the proof resulted that Avogadro's principle, first only speculatively suspected, must necessarily be correct if the atomic hypothesis proves at all valid.

The discovery of the energy principle (Mayer, Joule) further strengthened confidence in atomism. It was recognized that mechanical energy, which disappears through friction or the like, reappears in the quantity of heat produced — in such a way that a definite amount of consumed mechanical energy corresponds to a definite number of calories of heat. Conversely, according to the same conversion relation, in steam engines quantities of heat are transformed again into mechanical kinetic energy.

Conversion of Mechanical Energy

This conversion of mechanical energy into heat and its converse can be considered quite consistent, in the sense of "macrophysics". Whereupon one must establish that energy, which has proved indestructible, can assume both the form of a quantity of heat and of mechanical energy. The "macrophysical", "phenomenologic" heat theory thus arrived at is entirely sufficient for answering all the questions encountered in connection with heat engines, and with heat conversion in chemical processes.

But further thought is suggested to give the process of the conversion of mechanical energy into heat a clear interpretation in the light of the atomic theory. In a gas, as was represented above, the concepts of heat and temperature are not applicable to the individual molecules, but only to the gas mass as a whole. In an individual molecule only the kinetic energy, *i.e.*, its velocity of motion, or at best also its rotation, can be changed.

Thus, if energy is added to a gas mass, which appears macrophysically as added heat, it can only mean that the average energy of motion of the gas molecules has increased. The exact determination of the connection between the average energy of motion of the gas molecules and the temperature of the gas is a part of the theoretical investigations on the "kinetic theory of gases".

Similarly, in a solid body, a crystal perhaps, the temperature and heat energy contained within it are conceived of as mechanical energy of its molecules or atoms. Although in their regular arrangement in layers no one atom can leave its place, fine vibrating motions within the crystal — possibly similar to the fine vibration of a heavily travelled steel bridge — are still possible.

If these vibrations within the crystal are so small that they are not perceptible to us as mechanical motion, we do note the energy of these motions — as heat content which manifests itself in the temperature of the body. (If the atoms contained in each small piece of the crystal move by very small amounts in random directions, the whole crystal appears motionless to our macrophysical senses.)

Apropos of such thought processes was Boltzmann's atomistic interpretation of a remarkable regularity in heat phenomena. If we imagine that a planet's motion is interrupted and the planet then is induced to move in the opposite direction with the same velocity, it will retrace its entire elliptical path in the opposite direction after the reversal. This is an example of the fact that all purely mechanical motions are, as we say, reversible. The upward motion of a stone, thrown from the earth, (in the ideal case) until its-point of reversal is an exact temporal mirror image of the subsequent downward motion.

Similar reversibility — which can also be denoted as a physical "symmetry of th e positive and negative time orientation" — exists in all purely electromagnetic properties. But when we move a body over a table surface against frictional forces and heat is generated thereby, there is no reason to expect that in moving the body backwards along the same line the converse is true — that the heat energy will change back to energy of mechanical motion. Or take an electric current which flows through a wire and produces heat — when it flows through the wire in the opposite direction it again produces heat; the heat energy does not change back into electrical. These are examples of irreversible processes.

Another example of such a process is the mixing of two liquids; in most cases these intermix by themselves, but this process never runs backward of itself. A further example of importance to us is the following: if a gas filled vessel is introduced into an evacuated one and is opened there, the gas diffuses throughout the entire available volume. It is not impossible to restore the original conditions; the larger vessel can be evacuated again by pumping and the gas can be compressed into the smaller one.

But this does not constitute an exact reversal — a temporal reflected image — of the first process; and it can be considered that for the restoration of original conditions after any irreversible process a definite "compensation" must be ncluded (Clausius). In the steam engine, which undertakes to convert quantities of heat into mechanical energy, there does not occur a simple reversal of the process of the production of heat through friction, but each steam engine functions according to a scheme similar to the following: a certain amount of heat is removed from a heat container maintained at a high temperature.

A fraction of this amount is changed into mechanical work, while the remainder is conducted into a heat reservoir of lower temperature. As "compensation", therefore, for the conversion of heat into work there occurs the transfer of a quantity of heat from a hotter body to a colder one — *i.e.*, therefore a process which nature could perform "irreversibly" itself, since in heat conduction the heat always flows from higher to lower temperature. The occurrence of such "irreversible" processes among heat phenomena posed a very difficult problem for the "kinetic theory of heat" — tracing back the laws of heat to the mechanics of atoms. For purely mechanical processes were shown to be always reversible, which makes it appear impossible to trace these irreversible heat phenomena back to mechanical properties.

The solution of this difficulty, achieved through Boltzmann's perspicacity, runs as follows: in an aggregation of many similar particles, such as are attributed to every physical system by the atomic theory, there is no point to pursuing the motion of each individual atom or molecule; it is only important to us to observe the average statistical behaviour of large numbers of atoms. Thus statistical concepts are necessarily introduced into the consideration and we must determine which (defined in a rough statistical sense) events are to be considered as especially probable (thus as occurring very frequently) in comparison with other conceivable processes.

For our example this signifies that when the small gas filled vessel is opened in the large empty one "it is extremely probable" that the gas will stream out and distribute itself uniformly in the large vessel. Strictly speaking it cannot be definitely known that this will happen; indeed, to view the processes of motion in the gas mass with the same complete certainty achieved for planetary motions one must know quite exactly the position and velocity of each individual molecule at the beginning of the experiment. The fact that, instead of this, we simply denoted the initial condition of the experiment statistically stipulates that we can not predict its further progress with complete certainty, but only with very great probability.

The reversibility of pure physical processes, which according to the atomistic conception is the basis for macrophysical heat phenomena, depends upon the existence of an exact reversal, an exact temporal reflected image for every process — whereby it can happen that the gas mass distributed large vessel contracts into the smaller one, in an exact reversal of the normal process. No one can guarantee that this is completely out of the question; but we can show that such an event is extremely improbable — mathematically its probability is expressed by an infinitesimally small number. Thus it becomes clear that despite the fundamental reversibility of atomistic elementary- processes, yet, on a large scale, in macrophysical events, practically irreversible phenomena take place. Another instructive example of an irreversible process is the mixing of two gases or liquids with each other. According to the kinetic- atomic theory this problem is somewhat similar to the one wherein two types of balls — red and white ones for instance — are thoroughly shaken up in a sack.

If the red and white spheres are carefully separated initially, they will become completely mixed up after further agitation. If, then, we continue to shake, it is possible in principle that the original separation of red and white balls will be restored — but practically this possibility is of little moment since it would require shaking for astronomically long periods of time before this "spontaneous separation" could be expected with appreciable probability.

A quantitative measure for the irreversibility of a thermodynamic process (from a purely macrophysical standpoint) can be specified in the form of "entropy" — a quantity which always increases to a maximum in irreversible processes and does not decrease again (Clausius). According to Boltzmann this entropy can now be defined through the atomic concept as a measure of the disorder in an atomic aggregate.

AN ATOMIC ORBITALS

An atomic orbital is a mathematical function that describes the wave-like behaviour of either one electron or a pair of electrons in an atom. This function can be used to calculate the probability of finding any electron of an atom in any specific region around the atom's nucleus. The term may also refer to the physical region where the electron can be calculated to be, as defined by the particular mathematical form of the orbital. Atomic orbitals are typically categorized by *n*, *l*, and *m* quantum numbers, which correspond to the electron's energy, angular momentum, and an angular momentum vector component, respectively.

Each orbital is defined by a different set of quantum numbers and contains a maximum of two electrons. The simple names s orbital, p orbital, d orbital and f orbital refer to orbitals with angular momentum quantum number $l = 0, 1, 2$ and 3 respectively. These names indicate the orbital shape and are used to describe the electron configurations. They are derived from the characteristics of their spectroscopic lines: sharp, principal, diffuse, and fundamental, the rest being named in alphabetical order (omitting j).

Atomic orbitals are the basic building blocks of the atomic orbital model (alternatively known as the electron cloud or wave mechanics model), a modern framework for visualizing the microscopic behaviour of electrons in matter. In this model the electron cloud of a multi-electron atom may be seen as being built up (in approximation) in an electron configuration that is a product of simpler hydrogen-like atomic orbitals. The repeating *periodicity* of the blocks of 2, 6, 10, and 14 elements within sections of the periodic table arises naturally from the total number of electrons which occupy a complete set of s, p, d and f atomic orbitals, respectively. Atomic orbitals are in a way that makes them understandable for introductory courses and its equivalents. It explores s and p orbitals in some detail, including their shapes and energies. d orbitals are described only in terms of their energy, and f orbitals only get a passing mention.

ORBITALS AND ORBITS

When the a planet moves around the sun, you can plot a definite path for it which is called an orbit. A simple view of the atom looks similar and you may have pictured the electrons as orbiting around the nucleus. The truth is different, and electrons in fact inhabit regions of space known as *orbitals*. Orbits and orbitals sound similar, but they have quite different meanings. It is essential that you understand the difference between them.

The *Heisenberg Uncertainty Principle* says - loosely - that you can't know with certainty both where an electron is and where it's going next. It is impossible to define with absolute precision, at the same time, both the position and the momentum of an electron. That makes it impossible to plot an orbit for an electron around a nucleus. Is this a big problem? No. If something is impossible, you have to accept it and find a way around it.

Hydrogen's Electron - the 1s Orbital

Suppose you had a single hydrogen atom and at a particular instant plotted the position of the one electron. Soon afterwards, you do the same thing, and find that it is in a new position. You have no idea how it got from the first place to the second. In the hydrogen case, the electron can be found anywhere within a spherical space surrounding the nucleus.

The diagram shows a *cross-section* through this spherical space .95 per cent of the time (or any other percentage you choose), the electron will be found within a fairly easily defined region of space quite close to the nucleus. Such a region of space is called an *orbital.* You can think of an orbital as being the region of space in which the electron lives. What is the electron doing in the orbital? We don't know, we can't know, and so we just ignore the problem! All you can say is that if an electron is in a particular orbital it will have a particular definable energy. Each orbital has a name. The orbital occupied by the hydrogen electron is called a *1s orbital.*

The *"1"* represents the fact that the orbital is in the energy level closest to the nucleus. The *"s"* tells you about the shape of the orbital. s orbitals are spherically symmetric around the nucleus - in each case, like a hollow ball made of rather chunky material with the nucleus at its centre. The orbital on the left is a *2s orbital.* This is similar to a 1s orbital except that the region where there is the greatest chance of finding the electron is further from the nucleus - this is an orbital at the second energy level. If you look carefully, you will notice that there is another region of slightly higher electron density (where the dots are thicker) nearer the nucleus. ("Electron density" is another way of talking about how likely you are to find an electron at a particular place.) 2s (and 3s, 4s, etc.) electrons spend some of their time closer to the nucleus than you might expect. The effect of this is to slightly reduce the energy of electrons in s orbitals. The nearer the nucleus the electrons get, the lower their energy. 3s, 4s (etc) orbitals get progressively further from the nucleus.

p Orbitals

Not all electrons inhabit s orbitals (in fact, very few electrons live in s orbitals). At the first energy level, the only orbital available to electrons is the 1s orbital, but at the second level, as well as a 2s orbital, there are also orbitals called *2p orbitals.* A p orbital is rather like 2 identical balloons tied together at the nucleus. The diagram on the right is a cross-section through that 3- dimensional region of space. Once again, the orbital shows where there is a 95 per cent chance of finding a particular electron. Unlike an s orbital, a p orbital points in a particular direction.

At any one energy level it is possible to have three absolutely equivalent p orbitals pointing mutually at right angles to each other. These are arbitrarily given the symbols p_x, p_y and p_z. This is simply for convenience - what you might think of as the *x, y* or *z* direction changes constantly as the atom tumbles in space. The porbitals at the second energy level are called $2p_x$, $2p_y$ and $2p_z$. There are similar orbitals at subsequent levels - $3p_x$, $3p_y$, $3p_z$, $4p_x$, $4p_y$, $4p_z$ and so on. All levels except for the first level have *p* orbitals. At the higher levels the lobes get more elongated, with the most likely place to find the electron more distant from the nucleus.

d and f Orbitals

In addition to s and p orbitals, there are two other sets of orbitals which become available for electrons to inhabit at higher energy levels. At the third level, there is a set of five d orbitals (with complicated shapes and names) as well as the 3s and 3p orbitals ($3p_x$, $3p_y$, $3p_z$). At the third level there are a total of nine orbitals altogether.

At the fourth level, as well the 4s and 4p and 4d orbitals there are an additional seven f orbitals - 16 orbitals in all. s, p, d and f orbitals are then available at all higher energy levels as well. For the moment, you need to be aware that there are sets of five d orbitals at levels from the third level upwards, but you probably won't be expected to draw them or name them. Apart from a passing reference, you won't come across f orbitals at all.

Fitting Electrons into Orbitals

You can think of an atom as a very bizarre house (like an inverted pyramid!) - with the nucleus living on the ground floor, and then various rooms (orbitals) on the higher floors occupied by the electrons. On the first floor there is only 1 room (the 1s orbital); on the second floor there are 4 rooms (the 2s, $2p_x$, $2p_y$ and $2p_z$ orbitals); on the third floor there are 9 rooms (one 3s orbital, three 3p orbitals and five 3d orbitals); and so on. But the rooms aren't very big... Each orbital can only hold 2 electrons. A convenient way of showing the orbitals that the electrons live in is to draw "electrons-in-boxes".

"Electrons-in-Boxes"

Orbitals can be represented as boxes with the electrons in them shown as arrows. Often an up-arrow and a down-arrow are used to show that the electrons are in some way different.

A 1s orbital holding 2 electrons would be drawn as shown on the right, but it can be written even more quickly as $1s^2$. This is read as "one s two" - not as "one s squared". You mustn't confuse the two numbers in this notation:

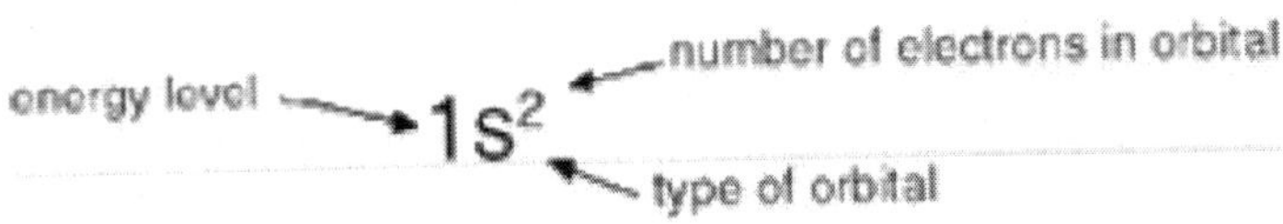

The Order of Filling Orbitals

Electrons fill low energy orbitals (closer to the nucleus) before they fill higher energy ones. Where there is a choice between orbitals of equal energy, they fill the orbitals singly as far as possible. This filling of orbitals singly where possible is known as *Hund's rule*. It only applies where the orbitals have exactly the same energies (as with p orbitals, for example), and helps to minimise the repulsions between electrons and so makes the atom more stable.

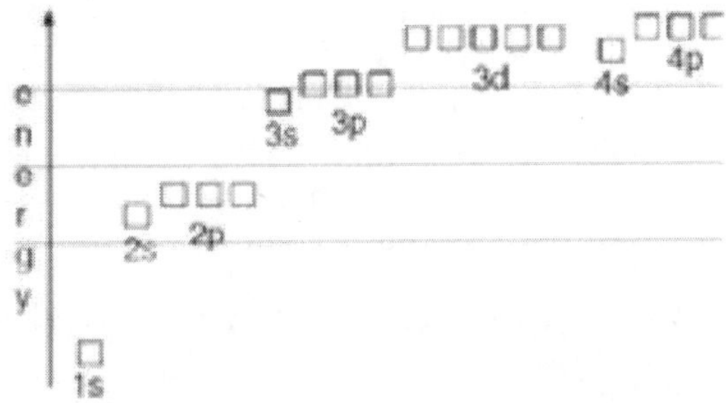

The diagram (not to scale) summarises the energies of the orbitals up to the 4p level.

orbitals at the same energy level, so the s orbital always fills with electrons before the corresponding p orbitals. The real oddity is the position of the 3d orbitals. They are at a slightly higher level than the 4s - and so it is the 4s orbital which will fill first, followed by all the 3d orbitals and then the 4p orbitals. Similar confusion occurs at higher levels, with so much overlap between the energy levels that the 4f orbitals don't fill until after the 6s, for example.

SPECTRUM

The first person to realise that white light was made up of the colours of the rainbow was Isaac Newton, who in 1666 passed sunlight through a narrow slit, then a prism, to project the coloured spectrum on to a wall.

This effect had been noticed previously, of course, not least in the sky, but previous attempts to explain it, by Descartes and others, had suggested that the white light became coloured when it was refracted, the colour depending on the angle of refraction. Newton clarified the situation by using a second prism to reconstitute the white light, making much more plausible the idea that the white light was composed of the separate colours.

He then took a monochromatic component from the spectrum generated by one prism and passed it through a second prism, establishing that no further colours were generated. That is, light of a single colour did not change colour on refraction. He concluded that white light was made up of all the colours of the rainbow, and that on passing through a prism, these different colours were refracted through slightly different angles, thus separating them into the observed spectrum.

In 1752, the Scottish physicist Thomas Melvill discovered that putting different substances in flames, and passing the light through a prism, gave differently patterned spectra. Ordinary table salt, for example, generated a "bright yellow". Furthermore, not all the colours of the rainbow appeared - there were dark gaps in the spectrum, in fact for some materials there were just a few patches of light. By the 1820's,

Herschel had recognized that spectra provided an excellent way to detect and identify small quantities of an element in a powder put into a flame.

Meanwhile, the white light of the sun was coming in for more detailed scrutiny. In 1802, William Wollaston in England had discovered that in fact the solar spectrum itself had tiny gaps - there were many thin dark lines in the rainbow of colours. These were investigated much more systematically by Joseph von Fraunhofer, beginning in 1814. He increased the dispersion by using more than one prism. He found an "almost countless number" of lines. He labelled the strongest dark lines A, B, C, D, etc.

FRAUNHOFER'S DARK LINES AND MELVILL'S BRIGHT LINES

In 1849, Foucault of speed of light and pendulum fame examined the spectrum of light from a voltaic arc between carbon poles. He saw a bright double yellow line at exactly the same wavelength as Fraunhofer's dark D line in the solar spectrum. Investigating further, Foucault passed the sun's light through the arc, then through a prism. He observed that the D lines in the spectrum were even darker than usual. After testing with other sources, he concluded that the arc, which *emitted* light at the D line frequency, would also *absorb* light from another source at that frequency. This discovery did not surprise Sir George Stokes in Cambridge. He pointed out that any mechanical system with a natural frequency of oscillation will emit at that frequency if disturbed, but will also absorb most readily at that frequency from incoming disturbances, the phenomenon of resonance.

The spectrum of *hydrogen*, which turned out to be crucial in providing the first insight into atomic structure over half a century later, was first observed by And ers Angstrom in Uppsala, Sweden, in 1853. His communication was translated into English in 1855. Angstrom, the son of a country minister, was a reserved person, not interested in the social life that centred around the court. Consequently, it was many years before his achievements were recognized, at home or abroad. Meanwhile, in Freeport, Pennsylvania, in 1855, David Alter described the spectrum of hydrogen and other gases. In the 1840's, Alter had started the first commercial production of bromine from brines.

BUNSEN AND KIRCHHOFF

The first really systematic investigation of spectra was that of Bunsen and Kirchhoff, in Heidelberg, between 1855 and 1863. They used several techniques. For one thing, they introduced various salts into — what else? — the flame of a Bunsen burner. This was a very effective way of viewing spectra, because the Bunsen burner flame itself gave out practically no light.

They also used the cooler flame of alcohol burning mixed with water to generate a vapour to study absorption spectra. Finally, they studied the spectra of electric arcs between electrodes of different materials. Using iron electrodes gave a spectrum that coincided with dark lines in the sun's spectrum. Copper electrodes did not. They concluded that the sun's atmosphere contained iron, but not much copper, and that, they said, seemed very plausible since there is so much iron in the earth, and in meteors. (*Cautionary note to philosophers*: In 1835, the French philosopher Auguste Comte (the founder of positivism) wrote: "...Our knowledge concerning the gaseous envelopes [of stars] is necessarily limited to their existence, size ... and refractive power, we shall not at all be able to determine their chemical composition or even their density ... I regard any notion concerning the true mean temperature of the various stars as forever effort, even by modern standards. They determined thousands of spectral lines, each to an accuracy of one part in ten thousand.

They spectroscopically discovered new elements: rubidium and cesium. Their method was used to find fifteen more new elements before the end of the century. In 1869, Joseph Lockyer studied the spectra of solar prominences (in eclipses). He found the spectra to be slightly Doppler shifted, so was able to deduce the speeds of the gases whirling around the sunspots. He also found a spectrum never seen before, and conjectured that it came from a new element he named Helium.

In fact, helium was later discovered on earth in 1895, by Ramsay. At that time, it had just become evident that there was an inert component, argon, in the earth's atmosphere. Earlier, an inert gas had been observed to emanate from uranium salts when they were heated. Ramsay assumed this would be the same gas, but decided to check. On heating uranium salts and performing a spectral analysis of the emitted gas, much to his surprise he found it to be helium. *The Balmer Series*: It is clear from the above that a tremendous amount of scientific progress was made using spectral lines, yet no-one had the slightest idea why atoms emitted at the frequencies they did. It was appreciated that spectra implied that atoms had structure. In 1852, Stokes had stated that probably the vibrations that produced light were vibrations among the constituent parts of molecules a term which also included atoms at that time and in 1875 Maxwell, in enumerating properties atoms must have, included the capability of internal motion or vibration. This worried Maxwell, though.

As he said, the spectroscopic evidence forces the conclusion that the atom is quite complex, with many internal degrees of freedom. Yet apparently all these modes of vibration, or almost all of them, are not excited by heat, since if they were this extra capacity of the atom to absorb energy would be reflected in its specific heat.

Obviously, if any pattern could be discerned in the spectral lines for an atom, that might be a clue as to the internal structure of the atom. One might be able to build a model. A great deal of effort went into analysing the spectral data from the 1860's on. The big breakthrough was made by Johann Balmer, a math and Latin teacher at a girls' school in Basel, Switzerland. Balmer had done no physics before, and made his great discovery when he was almost sixty. He decided that the most likely atom to show simple spectral patterns was the lightest atom, hydrogen. Angstrom had measured the four visible spectral lines to have wavelengths 6562.10, 4860.74, 4340.1 and 4101.2 in Angstrom units (10-10 metres). Balmer concentrated on just these four numbers, and found they were given by the formula:

$$\lambda = b\left(\frac{n^2}{n^2 - 4}\right)$$

where $b = 3645.6$ Angstroms, and $n = 3, 4, 5, 6$. Balmer suggested that there would be other lines - in the infrared - corresponding to $n = 7, 8$, etc., and in fact some of them had already been observed, unbeknownst to Balmer. He further conjectured that the 4 could be replaced by 9, 16, 25, … and this also turned out to be true - but these lines, further into the infrared, were not detected until the early twentieth century, along with the ultraviolet lines generated by replacing the 4 by 1.

It is instructive to write Balmer's general formula in terms of the inverse wavelength. This is called the wave number - the number of waves that fit in one unit of length.

$$\frac{1}{\lambda} = R_H\left(\frac{1}{n^2} - \frac{1}{m^2}\right)$$

where n, m are integers, and R_H is the Rydberg constant, 109,737 cm^{-1}.

This constant is named after the Swedish physicist Rydberg who (in 1888) presented a generalization of Balmer's formula, in which the integer n was replaced by n + constant, the constant being less than unity. Rydberg suggested that all atomic spectra formed families with this pattern. .He also said he was unaware of Balmer's work. It turns out that there *are* families of spectra following Rydberg's pattern, notably in the alkali metals, sodium, potassium, etc., but not with the precision the hydrogen atom lines fit the Balmer formula, and low values of n give lines that deviate considerably. (Modern footnote: atoms having spectral lines following Rydberg's formula are called *Rydberg atoms*.

Rydberg atoms have one electron orbiting at a much greater distance from the nucleus than the others. Consequently, Rydberg atoms can only survive in a gas at very low pressure, otherwise that outermost electron gets knocked off. Prof. Tom Gallagher in our Department is a world expert on these atoms, which have proven a rich source of information on the quantum mechanics of atomic structure.)

One pattern that was noticed in the spectra of *many* atoms is *Ritz' Combination Principle*: If for a given atom there are spectral lines at two wave numbers, there is sometimes another spectral line at the precise sum of those two wave numbers. It is easy to see from Balmer's formula that this is true for some pairs of lines in the hydrogen spectrum. It also turns out to be true for atoms where the spectral lines have no other discernible pattern.

FLOATING MAGNETS

In 1878, Alfred Mayer, at the University of Maryland, dreamed up a neat demonstration of how he imagined atoms might be arranged in molecules. He took a few equally magnetized needles and stuck them through corks so that they would float with their north poles all at the same height above the water, all repelling each other equally. He then held the south pole of a more powerful magnet some distance above the water, to attract the needles towards this central point. The idea was to see what equilibrium patterns the needles would form for different numbers of needles.

He found something remarkable - the needles liked to arrange themselves in shells. Three to five magnets just formed a triangle, square and pentagon in succession. but for six magnets, one went to the centre and the others formed a pentagon. For more magnets, an outer shell began to form. Kelvin's immediate response to Mayer's publication was that this should give some clues about the vortex atom. Apparently it didn't, but twenty-five years later it guided his thinking on a new model.

PLUM PUDDING

Kelvin, in 1903, proposed that the atom have the newly discovered electrons embedded somehow in a sphere of uniform positive charge, this sphere being the full size of the atom. (Of course, the sphere itself must be held together by unknown non-electrical forces - which is still true of the positive charge in our modern model of the atom.) This picture was taken up by J. J. Thomson too, and was dubbed the plum pudding model, after traditional English Christmas fare, a large round pudding (rich with suet) with raisins embedded in it.

In 1906, J. J. concluded from an analysis of the scattering of X-rays by gases and of absorption of beta-rays by solids, both of which he assumed were effected by electrns, that the number of electrons in an a tom was approximately equal to the atomic number. This led to a picture of electron arrangements in an atom reminiscent of Mayer's magnets. Perhaps by analysing possible modes of vibration of electrons in these configurations, the spectra could be calculated. The simplest case to consider was clearly hydrogen, now assumed (correctly) to contain just one electron.

How does an atom's colour depend on its size? By "colour" we mean here the spectral colours emitted when the atom is excited. In Thomson's plum pudding model, there is a clear relationship between the *size* of the pudding and the *frequency* at which the electron will oscillate, and hence presumably radiate, when excited. The two are related because the assumption is that the total positive charge - which is uniformly spread throughout the sphere — is just equal to the electron's negative charge.

At rest in its lowest state, the electron just sits in the middle of this sphere of charge. When bumped somehow, it will oscillate about that point. If the electron is at distance x from the centre, it will feel a restoring force towards the centre equal to the attraction from that part of the positive charge it is "outside" of - that is, the charge within a sphere of radius x about the centre. Therefore, the *larger* the whole atom — the pudding - the more thinly spread the positive charge is, and the *smaller* the amount of charge within the small sphere of radius x that is attracting the electron back towards the centre. So, the bigger the atom is, the slower the electron's

strength of attraction $e^2 / 4\pi\varepsilon_o$. Together with h, these parametres determine a length.

It should be mentioned that this assumption explained more than the size of the hydrogen atom. It was believed at the time that in the higher atoms, the electrons formed rings, thought to lie one outside the other, and various stability arguments indicated that there couldn't be more than seven electrons in a ring. The length scale above, $4\pi\varepsilon_o h^2 / \pi^2 me^2$, would *decrease* for larger atoms, with e^2 replaced by Ze^2 essentially, for nuclear charge Z. Thus as the number of rings increased, the size of the rings would decrease, explaining the observed approximate periodicity in atomic volume with atomic number.

Also, in 1911 Richard Whiddington in Cambridge had found that to cause a substance having atomic number A to emit characte ristic x-rays by bombarding it with electrons, it was necessary to use electrons of speed approximately $A \times 106$ metres per second. Any substance on being bombarded with sufficiently fast electrons emits a continuum of x-ray frequencies up to a maximum frequency f given by hf = kinetic energy of electron, *plus* some sharply defined lines — x-rays at a particular frequency, which does not change as the electron speed is further increased.

The frequency corresponding to these lines was found to increase with atomic number. Applying his length scale argument to the innermost ring of an atom, Bohr found that an electron in that ring would have a speed proportional to the nuclear charge, and hence, at least approximately, to the atomic number. Furthermore, the predicted speed in orbit was of the same order as that of Whiddington's electrons.

SPECTRUM OF THE HYDROGEN ATOM

Since the hydrogen atom has only one orbital electron surrounding a nucleus consisting of a single proton, it has a particularly simple spectrum. In this experiment, part of this spectrum will be determined. The observed frequencies of the lines can then be compared with the values predicted by quantum mechanics.

THEORY

When a high-voltage discharge takes place in H_2 gas at low pressures, many molecules are dissociated into atoms by electron impact. These atoms are often produced in excited electronic states. Such H atoms in excited electronic states spontaneously undergo transitions to lower energy electronic states with the emission of readiation. Quantum mechanics (to a high detgree of approximation) gives an expression for the allowed electronic energy levels of an H atom in terms of a single (principle) quantum number, n

$$E_n = -\frac{\mu e^4}{8h^2 \in_0^2} \frac{1}{n^2};$$

$$n = 1, 2, 3 \ldots$$

where e is the charge on the electron, h is Planck's constant, and epsilon_0 is the permittitivity of vacuum. The reduced mass mu is given in terms of the mass of the electron m and the mass of the proton M by,

$$\mu = \frac{mM}{m+M}$$

There is no selection rule for changes in n; that is, transitions may occur between any of the levels. An energy-level diagram for the H atom. Transitions from upper energy levels to lower levels are shown by the vertical lines. The transitions form several series of lines depending on the quantum number of the lower state. Note that the energy levels converge towards a limit as $n \to$ infinity.

The shaded area above n = infinity indicates a continuum of energy states corresponding to complete separation of the proton and the electron (*i.e.* ionization). We shall be concerned with the Balmer series, the lines of which fall in the visible region of the spectrum. It is possible to predict the frequency of a transition from any upper state with quantum number n, to any lower state with quantum number n^2, since

$$\Delta E = E_{n_1} - E_{n_2} = hv = hc\overline{v}$$

where c is the speed of light in cm.s-1 units and $(\overline{v})$ is the frequency in wavenumber units of cm^{-1} (equal to the reciprocal of the wave length (λ) in centimetres).

Thus:

$$\overline{v} = -\frac{\mu e^4}{8h^2 \in_0^2}\left(\frac{1}{n_2^2} - \frac{1}{n_1^2}\right) = R\left(\frac{1}{n_2^2} - \frac{1}{n_1^2}\right)$$

The quantity R is called the Rydberg constant and has the calculated value 109 677.5805 cm^{-1} for the reduced mass of the H atom. For the Balmer series, $n^2 = 2$ and $n^1 = (3, 4, 5,...)$. The above equation predicts a series of lines in the 2 Balmer series that converge to a high-frequency limit at R / n^{2} wavenumbers.

Balmer series that converge to a high-frequency limit at R / n^2 wavenumbers.

You should probably scan this region in your experiment to see if you can directly detect the Ionization Potential of H in the $n = 2$ state.

METHOD

The frequency (or wavelength) of the Balmer lines can be determined experimentally by comparing the hydrogen spectrum with a reference spectrum for which the wavelengths are known. The iron, neon, or mercury spectra are common choices for reference spectra. (Such reference spectral lines must, of course, have been measured absolutely, as by an interferometric wavelenght measurement or by direct frequency counting.)

In this experiment, a Ne lamp will be used because the lines of the Ne spectrum can be identified quite easily, and Neon displays are common and available. In other work, the Fe spectrum is preferable, since there are many more lines. Sometimes, lamps are constructed with several emitters present, so that even more calibration points can be obtained. A wide variety of spectrometres and spectrographs (instruments that record the spectra on a photographic plate) can be used to study spectra in the visible and near ultraviolet region.

A description of such instruments is given in most modern Instrumental textbooks, and this material should be read as background for the present experiment. A schematic diagram for a typical grating spectrometre, which consists of a monochromator, light source and detector. In this experiment, we will be studying the line output of the source, so no absorption sample is present. The source should be mounted in such a way that the hydrogen discharge tube will illuminate the slit at the same time as the Neon reference lamp. For wavelengths shorter than about 300 to 350 nm, most glasses absorb light strongly but quartz optics may be used. Even Quartz begins to absorb light below about 190 nm. Other factors, such as detector response, also limit the accessible wavelight range of a given experimental setup.

Since the Balmer lines studied here are more and more closely spaced and have decreasing intensity as the short-wavelength limit is approached, both high resolution and high sensitivity are needed to observe lines with high n1 values. For this reason, one may want to increase the gain and decrease the slit width to obtain the part of the Balmer spectrum at short wavelength. The determination of the Rydberg convergence limits and the assignment of the states involved in the atomic energies is always easier with at high resolution.

SOURCES

The best Ne light source is a simple low-pressure neon lamp that operates from a small ballast transformer. Neon bulbs which operate at 110 VAC are common panel display lights. Many designs of hydrogen discharge tubes are available. A convenient commercial hydrogen source is a Geissler tube, since the pressure has been adjusted (0.1 to 0.5 Torr) to give strong Balmer lines. In addition, there is always a weak band spectrum due to molecular hydrogen; hydrogen atoms, formed by the electrode discharge, may recombine on the walls to give H_2 molecules in excited electronic states which will emit radiation.

Such a band spectrum should be so weak that it will not interfere with determining the Balmer lines. The Geissler discharge tube operates from a neon-sign transformer at about 5000 V and 15 mA; it should glow steadily with a bright red colour. Instructions for the operation of the spectrometre to be used will be given in the laboratory. The adjustable parametres that need to be considered include: The slit widths, the PMT Voltage (gain), the stripchart gain and chart speed, the monchromator scan range and speed, and the position of the lamps. (BTW: always scan spectra in the same direction to avoid errors due to backlash in the screw drive on the grating). In this particular experiment, a computer will be taking the place of a strip chart recorder.

In traditional experiments, a strip chart recorder (which moves a strip of paper at constant speed) would be use to record the data coming from the PMT as the monochromator scans wavelength at constant rate. The strip chart pen would move proportional to the signal coming from the PMT and the distance down the paper would be proportional to the wavelength. In this lab you will use a computer programme and an Analog to Digital card to measure the signal, replacing the mechanical stripchart with digital data acquisition. It is importa nt to note that the c omputer is not synch ronize d with the monochromator at the beginning of a scan except by your actions (just like the stripchart recorder).

The computer will record data at regular intervals when it is told to and will continue to record data until it is told to stop. It becomes important to use a method in which the signal versus time readings may be converted into a signal versus wavelength readings (a spectrum). Calibration of wavelengths is crucial. There are several methods by which you can do this. One method is to produce a 'marker' on the computer plot at know wavelength by briefly blocking the light coming into the monochromator, producing a spike in your data. It is also possible to turn on and off the 'signal simulator' (a box in the signal cable with a switch and knob), which will also produce a spike in the spectrum. With this, you can at least transfer are a ding from the monochromator scanbox to the compare plot.

PROCEDURE

First you must turn on all the instrumentation, including the various power supplies and the monochromator. Allow the monochromator to 'warm up'. On the desktop of the computer, open the programme called"A-D Converter." Go to "DAQ type" and select "DAS005" and exit. Go to "Printerport" and select address.

In the section labeled "Other Address" type in "3bc." Then click on "Find Port" until something appears. The click "Intialize," and click on 'okay' to acknowledge printer port intialisation. Go to File Load and load the file called "." After this is done, go to "View" and select "Trends." This enables you to see the signal coming from the PMT. Set the signal level to approx. 4000 by carefully twisting the knob on the 'signal simulator'. The A-D software should now be able to record the signal. It is now time to look at a signal.

Turn on the desired lamp and place it near the entrance slit to the monochromator. You can set the starting wavelength by pressing "STA POS" on the scan box, typing in the desired value, and hitting 'enter'. The ending wavelength can be set in the same manner using the "END POS" key. The scanning rate can be set hitting the "Rate" button, typing in '0.3' and hitting 'enter'. The monochromator will now scan at a rate of 0.3 nm/sec. While still in the View ® Trends menuin the computer, hit Set / Sca n on t h e monochromator. This will move the monochromator to the start position.

By hitting Set/Scan a second time, the monochromator will begin to scan across the wavelengths. Observe the response. (Wavelengths for this test scan can be determined from the 'survey' scans included in this protocol. Make sure to observe all the bands in the convergence near 350 nm and all the prominent members of the Balmer series of H). If the signal is very saturated, decrease the PMT gain by reducing the High Voltage.

Once you are ready to begin recording data, go to "'Setup"' then go to "'Logging"' and then "'Parametres'." Go to 'Input file name' and type in the name of the file, making sure you save to correct drive. Every individual scan must have a new file name. Return to "Setup" then go to "Logging" and then choose the other option. This should bring up a screen with a checkbox in the lower left hand corner that says "Start Logging." A soon as you click in this check box the computer will begin to collect data at a rate of 5 samples/sec (how many nm/sample?). To stop logging data unclick the check box. It is possible to watch the data, by returning to the "View- Trends" window. Remember, the computer will continue to log data until you unclick the "Start Logging" checkbox.

In order to obtain good data it is possible to vary several settings. These include the slit width, the voltage setting on the High Voltage box, the base signal on the signal simulator (do not increase this above 4050 mV), and the scan rate. (What effect would each of these have on your signal?)

❐

6.

Quantum Dynamics

In quantum mechanics, particles are represented by matter waves, with the absolute square of the wave displacement yielding the probability of finding the particle. The behaviour of particles thus follows from the reflection, refraction, diffraction, and interference of the associated waves. The connection with Newtonian dynamics comes from tracing the trajectories of matter wave packets. Changes in the speed and direction of motion of these packets correspond to the accelerations of classical mechanics. When wavelengths are small compared to the natural length scale of the problem at hand, the wave packets can be made small, thus pinpointing the position of the associated particle, without generating excessive uncertainty in the particle's momentum. This is the geometrical optics limit of quantum mechanics.

RELATIVISTIC DYNAMICS

The first coherent statement of what physicists now call *relativity* was Galileo's observation almost four hundred years ago that if you were in a large closed room, you could not tell by observing how things move-living things, thrown things, dripping liquids-whether the room was at rest in a building, say, or below decks in a large ship moving with a steady velocity. More technically (but really saying the same thing!) we would put it that the laws of motion are the same in any inertial frame. That is, these laws really only describe relative positions and velocities. In particular, they do not single out a special inertial frame as the one that's "really at rest".

This was later all written down more formally, in terms of Galilean transformations. Using these simple linear equations, motion analysed in terms of positions and velocities in one inertial frame could be translated into any other. When, after Galileo, Newton wrote down his Three Laws of Motion, they were of course invariant under the Galilean transformations, and valid in any inertial frame.

A SYMMATRICAL SPACESHIP COLLESION

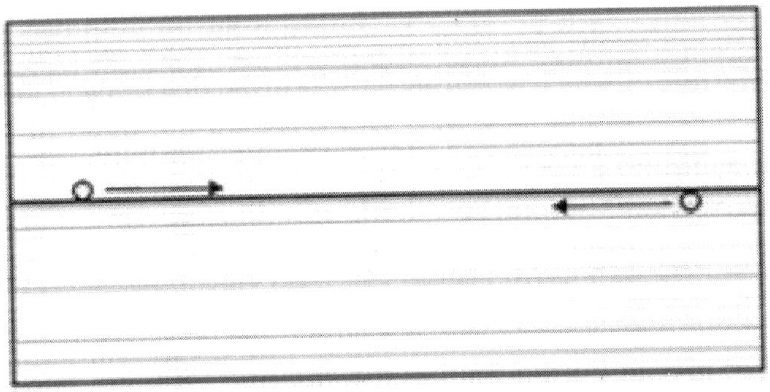

Balls on pool table moving towards glancing collision

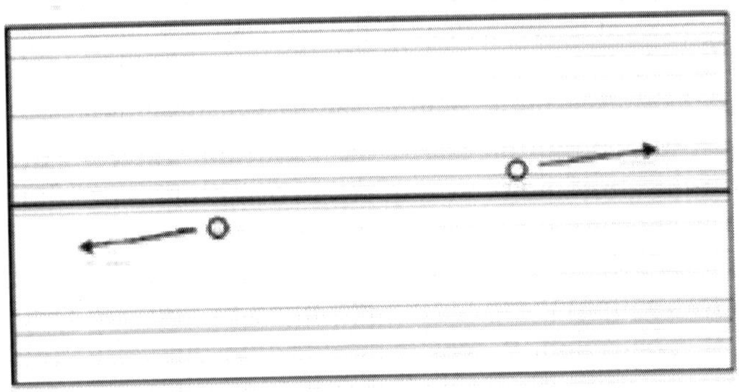

Motion of ball of Table after collesion

Now let us repeat the exercise on a grand scale. Suppose somewhere in space, far from any gravitational fields, we set out a string one million miles long. (It could be between our two clocks in the time dilation experiment).

This string corresponds to the chalk line on the pool table. Suppose now we have two identical spaceships approaching each other with equal and opposite velocities parallel to the string from the two ends of the string, aimed so that they suffer a slight glancing collision when they meet in the middle. It is evident from the symmetry of the situation that momentum is conserved in both directions. In particular, the rate at which one spaceship moves away from the string after the collision - its y-velocity - is equal and opposite to the rate at which the other one moves away from the string.

But now consider this collision as observed by someone in one of the spaceships, call it A. (Remember, momentum must be conserved in *all* inertial frames—they are all equivalent—there is nothing special about the frame in which the string is at rest.) Before the collision, he sees the string moving very fast by the window, say a few meters away. After the collision, he sees the string to be moving away, at, say, 15 meters per second. This is because spaceship A has picked up a velocity perpendicular to the string of 15 meters per second. Meanwhile, since this is a completely symmetrical situation, an observer on spaceship *B* would certainly deduce that her spaceship was moving away from the string at 15 meters per second as well.

JUST HOW SYMMETRICAL IS IT

The crucial question is: how fast does an observer in spaceship A see spaceship B to be moving away from the string? Let us suppose that relative to spaceship A, spaceship B is moving away (in the x-direction) at 0.6c.

First, recall that distances perpendicular to the direction of motion are not Lorentz contracted. Therefore, when the observer in spaceship *B* says she has moved 15 meters further away from the string in a one second interval, the observer watching this movement from spaceship *A* will agree on the 15 meters - but disagree on the one second! He will say her clocks run slow, so as measured by his clocks 1.25 seconds will have elapsed as she moves 15 meters in the *y*-direction.

It follows that, as a result of time dilation, this collision as viewed from spaceship *A* does *not* cause equal and opposite velocities for the two spaceships in the *y*-direction. Initially, both spaceships were moving parallel to the *x*-axis - there was zero momentum in the *y*-direction. Consider *y*-direction momentum conservation in the inertial frame in which *A* was initially at rest. An observer in that frame measuring *y*-velocities after the collision will find *A* to be moving at 15 meters per second, *B* to be moving at -0.8 x 15 meters per second in the *y*- direction. So how can we argue there is zero total momentum in the *y*-direction *after* the collision, when the identical spaceships do *not* have equal and opposite velocities?

EINSTEIN RESCUES MOMENTUM CONSERVATION

Einstein was so sure that momentum conservation must always hold that he rescued it with a bold hypothesis: the mass of an object must depend on its speed! In fact, the mass must increase with speed in just such a way as to cancel out the lower *y*-direction velocity resulting from time dilation. That is to say, if an object at rest has a mass m_0, moving at a speed v it must have mass

$$m = \frac{m_o}{\sqrt{1 - v^2 / c^2}}$$

to conserve *y*-direction momentum.

Note that this is an undetectably small effect at ordinary speeds, but as an object approaches the speed of light, the mass increases without limit! Of course, we have taken a very special case here: a particular kind of collision. The reader might well wonder if the same mass correction would work in other types of collision, for example a straight line collision in which a heavy object rear-ends a lighter object. The algebra is straightforward, if tedious, and it is found that this mass correction factor does indeed ensure momentum conservation for any collision in all inertial frames.

MASS REALLY *DOES INCREASE* WITH SPEED

Deciding that masses of objects must depend on speed like this seems a heavy price to pay to rescue conservation of momentum! However, it is a prediction that is not difficult to check by experiment. The first confirmation came in 1908, measuring the mass of fast electrons in a vacuum tube. In fact, the electrons in an old-fashioned colour TV tube are about half a percent heavier than electrons at rest, and this must be allowed for in calculating the magnetic fields used to guide them to the screen.

Much more dramatically, in modern particle accelerators very powerful electric fields are used to accelerate electrons, protons and other particles. It is found in practice that these particles become heavier and heavier as the speed of light is approached, and hence need greater and greater forces for further acceleration. Consequently, the speed of light is a natural absolute speed limit. Particles are accelerated to speeds where their mass is thousands of times greater than their mass measured at rest, usually called the "rest mass".

Warning: It should be mentioned that some people don't like the statement that mass increases with speed, they feel that the word "mass" should be restricted to the rest mass of an object, which we've called m_0. This difference of definition has no physical content, however—it's just a matter of taste. We would write momentum as $p = mv$, they would write our m_0 as m, and say the formula for momentum in their notation is, line.

$$p = mv / \sqrt{1 - v^2 / c^2}$$

Either way, a fast electron is that much harder to deflect from a straight

MASS AND ENERGY CONSERVATION:
KINETIC ENERGY AND MASS FOR VERY FAST PARTICLES

As everyone has heard, in special relativity mass and energy are not separately conserved, in certain situations mass m can be converted to energy $E = mc^2$. This equivalence is closely related to the mass increase with speed. Suppose a constant force F accelerates a particle of rest mass m_0 in a straight line. The work done by the force in accelerating the particle as it travels a distance d is Fd, and this work has given the particle kinetic energy. The elementary derivation of the kinetic energy $\frac{1}{2}mv^2$ of an ordinary *non-relativistic* (*i.e.* slow moving) object of mass m. Suppose it starts from rest. Then after time t, it has traveled distance $d = \frac{1}{2}at^2$, and $v = at$. From Newton's second law, $F = ma$, the work done by the force

$$fd = mad = \frac{1}{2}ma^2t^2 = \frac{1}{2}mv^2.$$

This won't work if the mass is varying, because Newton's Second Law isn't always $F = ma$, for variable mass it's

$$F = dp/dt,$$

force = rate of change of momentum, and if the mass changes the momentum changes, even at constant velocity.

An instructive extreme case is the kinetic energy of a particle travelling close to the speed of light, as particles do in accelerators. In this regime, the change of speed with increasing momentum is negligible! Instead,

$$F = \frac{dp}{dt} = \frac{d(mv)}{dt} \cong \frac{dm}{dt}c$$

where as usual c is the speed of light. This is what happens in a particle accelerator for a charged particle in a constant electric field, with $F = qE$.

Since the particle is moving at a speed very close to c, in time dt it will move cdt and the force will do work $Fcdt$. The equation above can be rewritten

$$Fcdt = (dm)c^2$$

So the energy dE expended by the accelerating force in the time dt yields an increase in mass, and $dE = (dm)c^2$. Provided the speed is close to c, this can of course be integrated to an excellent approximation, to relate a finite particle mass change to the energy expended in accelerating it.

CHEMICAL DYNAMICS

Our major concern has been whether or not a chemical reaction of interest may or may not occur under specified conditions of temperature, pressure, and concent ration. The answer to this question is in the realm of thermodynamics. If thermodynamics tells us that a reaction may occur, then we must be concerned with how rapidly and by what pathway it will occur, and how we may influence this.

The speed (rate) at which a reaction takes place is a matter of chemical dynamics, or alternately, chemical kinetics. Chemical dynamics (kinetics) is the study of the rates (speeds) and mechanisms (pathways) of chemical reactions. We have briefly encountered the rate concept before, when we discussed the manner in which vapour pressure changes with time, and w hen we pres ented dynamic equilibrium a s a situa tion of ba la nc ed, opposing rates.

Allthough chemical kinatics is a well-developed field experimentally, we are still a long way from the theoretical grasp of rate that we have of equilibrium. Thus our predictive ability in the realm of kinetics is very limited. Thermodynamics may predict that a reaction should proceed essentially completely. Whether it does so depends on the rate with which it occurs. For example, consider the reaction below at 298 K:

$$C(graph) + O_2(g) \rightarrow CO_2(g)\left[DG_R^0 = 394.4kJ\right]$$

Thermodynamics predicts that the reaction should go to completion. But carbon (coal) deposits exist in the atmosphere without noticeably reacting; and the "lead" in our pencils does not burst spontaneously into flame. The reaction is favoured, but is very slow at 298 K. We say that C(graph) is thermodynamically unstable in the presence of O_2(g) at 298 K, but is kinetically inert. As a second case, consider the acid-base neutralization reaction:

$$H_3O^+(aq) + OH^-(aq) \rightarrow 2H_2O(l)\left[DG_R^0 = 58.28KJ\right]$$

Thermodynamics predicts that this reaction, too, should go to completion. In fact, it occurs almost instantaneously; equilibrium is attained as rapidly as the reactants are mixed. Thus we say that H_3O^+ is both thermodynamically unstable and kinetically labile in the presence of OH–. Two sets of descriptive words are necessary to distinguish thermodynamics and kinetics: Thermodynamics: stable DG > 0; unstable DG < 0 Kinetics: inert (reaction is slow); labile (reaction is fast). As we have seen above, there is no necessary correlation between "stable" and "inert", or "unstable" and "labile". This lack of correlation reflects the fact that thermodynamics deals with values of state functions in various equilibrium states of a system; kinetics deals with the rate at which a path between states is traversed.

Reaction rates are extremely important in many ways in our lives. Sustained muscle action depends on the continuous supply of energy produced by glucose combustion in the mitochondria of cells. Energy production must be fairly rapid, otherwise sustained action is not possible. The extremely rapid (explosive) burning of gasoline in the internal combustion engine provides the motive power for automobiles. Food cooks (or, as the chemist would say, oxidizes) extremely slowly at room temperature, but in a relatively short time at more elevated temperature. Our first concern must be to discuss the manner in which reaction rates are expressed, and to introduce some of the terminology of chemical dynamics.

RATE

In general, a rate is a number expressing how much a quantity changes in a convenient interval of time. Thus an interest rate is the amount by which a sum of money increases in 1 year; travel rate in an automobile or airplane is the distance covered in 1 hour; heart rate is the number of heartbeats per minute; thus a rate is the change in a quantity per unit time:

Rate = Dquantity/Dt

Note that the units of rate must be "quantity/time, or "quantity-time–1". Thus interest rate has units of dollars/year; travel rate, miles/hour; and heart rate; beats/minute.

How are we to express the rate of a chemical reaction? To develop the answer to this question, we will begin with the simple chemical process in equation. The equation describes quantitatively the conversion of ozone to oxygen that occurs in the earth's upper atmosphere.

$$2O_3(g) \rightarrow 3O_2(g)$$

When this (or any other) reaction takes place, it is the amounts of reactants and products that change. As we know, amount can be expressed in either mass, moles, or molecules.

The coefficients in the chemical equation relate both molecules and moles of reactants and products but do not directly relate masses. So we will agree that our measure of amount should be either molecules or moles. Further, since work in the laboratory is carried out on the macroscopic scale, moles is the more suitable unit of amount.

So we might describe the rate of process either in terms of the number of moles of O_3 used or of O_2 produced per unit time. However, there is a major difficulty with this approach. The number of moles of ozone used will depend upon the volume of atmosphere that we choose to examine! We cannot specify a rate without also specifying the volume examined. Rather than use moles/ time, then, we choose to divide by the appropriate volume to give moles/ litre-time, or c onc entra tion/ time. Ra tes express ed in the se units a re independent of the actual amount of reaction carried out.

At this point then we can express the rate of reaction in either of two ways:

1. Rate = Moles O_3 used per litre per time = –D(conc O_3)/Dtime
2. Rate = Moles O_2 produced per litre per time = D(conc O_2)/Dtime

The units of the rate are then moles per litre per time, where the time unit should be the one most convenient for the reaction being measured. Chemists agree that the rate of a reaction should be expressed as a positive number.

As reaction proceeds, the concentration of ozone becomes smaller while that of oxygen becomes larger. Thus the change in ozone concentration, D(conc O_3), is a negative quantity. To produce a positive rate, it is necessary to place the minus sign preceding the D (conc) term, as has been done. In general, whenever the rate of a reaction is expressed in terms of the change in concentration of a reactant, D(conc reactant) is negative and the minus sign is necessary. The minus sign is unnecessary because Δ (conc O_2) is positive.

There is an additional difficulty, though. We would like to specify a single number representing the rate of a reaction under given conditions. However, the two expressions above give different numbers for the rate, because that 2 moles of ozone react, 3 moles of dioxygen appear). This can be fixed by normalizing the two expressions in with the stoichiometric coefficients. Dividing Δ (conc O_3) by the coefficient 2 gives the same number as dividing Δ (conc O_2) by the coefficient 3:

$$\text{Rate} = -1/2 * \Delta(\text{conc } O_3)/\text{Dtime} = 1/3 * \Delta(\text{conc } O_2)/\Delta\,\text{time}$$

Replacing "conc O_3" with the brackets to represent molar concentratration:

$$\text{Rate} = -1/2 * \Delta[O_3]/\Delta t = 1/3 * D\Delta[O_2]/\Delta t$$

The symbol, Δ, signifies a finite (*i.e.*, measurable) change in a quantity. In the limit as the length of the time interval, Dt, approaches zero, we replace D(conc) and Dt with the differential (infinitesimal) quantities D(conc) and dt to obtain the differential expression of rate.

$$\text{Rate} = -1/2\ \Delta[O_3]/dt = 1/3\ \Delta[O_2]/dt$$

The appropriate formulation for finite difference methods of analysis. For the completely general chemical reaction, proceeding from left to right from reactants, *A* and *B*, to products, D and *F*, the rate of reaction can be represented using any one

FACTORS AFFECTING REACTION RATE; THE RATE LAW

The rate of a chemical reaction depends on a number of factors, some of which are listed below:

- The molecular structure(s) of the reactant(s)
- The medium in which the reaction is carried out (for example, does the reaction occur in the gas phase, or does it occur with reactants dissolved in a solvent?)
- Temperature
- The presence or absence of a catalyst
- The concentrations of the reacting species

The dependence of rate on molecular structure is the most obvious but least understood factor. This complex matter is a focus of intense laboratory research. At present, it is not possible to summarize this dependence in a neat, clean package. In fact, this incompletely understood factor is what prevents predictive ability in kinetics.

For molecules to react, they must directly interact; to directly interact, they must collide. This leads us to expect that rate should be proportional to the frequency of molecular collisions. In turn, the frequency of collisions should depend on the number of molecule sperun it volume, which is the concentration. This concentration dependence of rate is indeed observed experimentally.

For most chemical reactions, the dependence of reaction rate on the concentrations of the reacting species is surprisingly simple. In almost all cases, the reaction rate is found to be proportional to the product of the concentrations of the reactants, each concentration raised to an integer power that is almost always 1 or 2. Thus for the general reaction, the rate law is expected to take the form:

$$\text{Rate} = -1/a \; \Delta \, [A]/dt = k\,[A]^x\,[B]^y$$

Equation is called the differential rate law, because it involves the ratio of differentials (derivative), D$[A]/dt$. The exponents, x and y, are integers, and for most reactions are either 1 or 2.

It must be stressed that in most cases x and y are different from the stoichiometric coefficients, a and b; it is not valid to write the rate law for a reaction by raising the concentrations to the stoichiometric powers, unless the reaction is an elementary process.

We will have more to say about elementary processes in a subsequent section. The integer, x, is called the order of the reaction in reactant A; y is the order in B. The sum, $x + y$, is called the overall reaction order.

The reaction orders must be determined by experiment; they can not be written down simply by looking at the reaction stoichiometry. To illustrate this statement for a specific case, the reaction below, involving 31 moles of reactants, has the accompanying rate law. This very complex overall reaction has a simple rate law, which is first order in each reactant, and second order overall.

$$30C_2H_5OH + B_{10}H_{14} \rightarrow 10B(OC_2H_5)_3 + 22H_2(g)$$

$$Rate = k[C_2H_5OH][B_{10}H_{14}]$$

Finally, k (and in the expression just above) is the proportionality constant relating Rate to the concentration product. It is called the rate constant for the reaction. The rate constant contains implicit within it information about the dependence of rate on molecular structure, reaction medium, and temperature. A word about the units of the rate constant is in order.

Generally, k has whatever units are required to give the rate units of concentration/time. Example.shows how units for the rate constant may be determined for various rate laws.

Example: For each rate law below, state the order of the reaction in each reactant, and determine the units of the rate constant, k.

- -d[A]/dt = k
- -d[A]/dt = k[A]
- -d[A]/dt = $k[A]^2$
- -d[A]/dt = k[A][B]
- -d[A]/dt = $k[A]^2[B]$

Solution:

Rate Law	Reaction Order	Units of k
-d[A]/dt = k	Zero order (no concentration dependence)	k has units of rate, Moles L^{-1} $time^{-1}$
-d[A]/dt = k[A]	First order in A	Moles/L-time = k(moles/L) so k has units of time-1
-d[A]/dt = $k[A]^2$	Second order in A	Moles/L-time = $k(moles/L)^2$ so k has units L moles-1 time-1
-d[A]/dt = k[A][B]	First order in A First order in B Second order overall	Same as for the previous case
-d[A]/dt = $k[A]^2[B]$	Second order in A First order in B third order overall	Moles/L-time = $k(Moles/L)^3$ so k has units L^2moles^{-2} $time^{-1}$

Some Specific Examples of Rate Laws. Zero-order Reactions. Chemical reactions having overall order zero are rare. However, evaporation of a pure liquid, is an example of a physical process with a zero-order rate law. Recall that the rate of evaporation of a liquid having exposed surface area, A, is just,

Rate = k*A

As long as the exposed surface area remains constant, we may rewrite this as,

Rate = k', where k' = k*A

Since this rate is independent of concentration of liquid, the process is zero order in liquid.

First-Order Reactions

Reactions and physical processes exhibiting first order rate laws are very c om mon. Conden sation is a first order ph ysical proces s. The rate of condensation of vapour to liquid in a closed system is given by

$$Rate = k^*P_{vap}$$

where P_{vap} is the vapour pressure of the substance. However, P is related to concentration in moles/L by the ideal gas law, leading to,

$$Rate = -d[vapour]/dt = k^*[vapour]$$

Molecular rearrangements (often called isomerization reactions) frequently follow first order rate laws. Vision is critically dependent upon cis-retinal $\rightarrow$ trans-retinal where cis- and trans- retinal contain the same atoms, bonded together in the same way, but with somewhat different spatial arrangements. The rate of conversion of cis- to trans- form is given.

$$-d[cis]/dt = k[cis]$$

Similarly, trans- converts to cis- by a first order rate law with a different rate constant. Finally, nuclear decay processes are first order. Thus ^{238}U decays according to equation, which is governed by the rate law.

$$^{238}U \rightarrow {}^{234}TH + {}^{4}He$$

$$Rate = -d(mass\ ^{238}U)/dt = k\ (mass\ ^{238}U)$$

The rate constant for this process has the extremely small value, 1.54×10^{-10} year^{-1}.

The rate constant, k, for a first-order process has units of time^{-1}. It's reciprocal, 1/k, thus has units of time. The quantity, 1/k, is given the symbol t, and is called the time constant for the reaction.

This is a measure of the time required for the reaction to take place. The time constant for decay of ^{238}U is $1/k = 6.49 \times 10^9$ years. The fact that this time constant is so large is the reason why ^{238}U is still found in the earth's crust, which is estimated to be about 4.5 billion years old.

Example: The time constant for the radioactive isotope, ^{226}Ra, is 2338 years. Calculate the number of atoms of ^{226}Ra that decay each minute in a 2.00 g sample of ^{226}Ra.

Solution: Nuclear decay is a first order process. We can therefore write the following rate law for decay of ^{226}Ra:

$$Rate = -d(^{226}Ra\ atoms)/dt = k(Number\ of\ ^{226}Ra\ atoms)$$

$$-dN/dt = kN$$

where N is the number of ^{226}Ra atoms in the sample, and $k = 1/t = 4.28 \times 10^{-4}$ year^{-1}.

$N = (2.00g/226.025\ g/mole) * 6.023 \times 10^{23}$ atoms/mole $= 5.33 \times 10^{21}$ atoms.

$-dN/dt = (4.28 \times 10\text{-}4\ year^{-1})(5.33 \times 10^{21}\ atoms) = 2.28 \times 10^{18}\ atoms/year$

There are 5.256×10^5 minutes in 1 year. Therefore,

$-dN/dt = (2.28 \times 10^{18} atoms/year)/(5.256 \times 105\ minutes/year)$

$= 4.34 \times 10^{12}\ atoms/minute$

Second Order Processes

Many reactions exhibit second-order rate laws, either simple (second-order in a single reactant) or mixed (first-order in each of two reactants). Here we consider two specific reaction types that exhibit second-order kinetics. Dimerization reactions, in which two molecules unite to form a larger molecule, obey simple second-order kinetics.

The dimerization of NO_2 is shown below:

$$2NO_2(g) \rightarrow N_2O_4(g)$$

The rate law:

$$-d[NO_2]/dt = k[NO_2]^2$$

where the second order rate constant is — $M^{-1}s^{-1}$ at oC. Other examples of dimerization processes are the formation of a C^{12} molecule from two C^1 atoms and the formation of one molecule of C_4H_8 from two molecules of C_2H_4. In general, a reaction of the general type is expected to obey simple second-order kinetics :

$$2A \rightarrow A_2$$

Adduct formation processes, such as Lewis acid-base reactions and the formation of complexes involving enzymes and their substrates, are also second-order processes. Most such processes occur very rapidly. The role of enzyme-substrate complexes in enzyme catalysis will be discussed in a subsequent section.

"Net" Reaction Rate; A Prelude to the Next Section

We discuss experimental approaches to the determination of rate laws for reactions. To preface that discussion, we briefly discuss here the concept of net reaction rate. Except in some cases, we can experimentally measure only the net rate of a reaction, where the net rate is defined as:

Net reaction rate = rate of forward reaction – rate of reverse reaction

$$Rate_{net} = Rate_{forward} - Rate_{reverse}$$

The general reaction is reproduced below, this time acknowledging that in general the reverse process also occurs:

$$aA + bB \Leftrightarrow dD + fF$$

For this reaction we expect the forward rate to depend only on the concentrations of reactants, A and B, and the reverse rate to depend only on the concentrations of products, D and F:

$$Rate_{forward} = k_f[A]^x[B]^y$$
$$Rate_{reverse} = k_r[D]^p[F]^q$$

The net rate is therefore expected to be a potentially complex function of the concentrations of reactants AND products. The experimentally determined rate constant is a composite function of the forward and reverse rate constants, k_f and k_r.

$$\text{Rate}_{net} = k_f[A]^x[B]^y - k_r[D]^p[F]^q$$

Once reaction reaches equilibrium, the net rate will of course be zero. Applying this idea to, our familiar criterion for dynamic equilibrium:

$$\text{Rate}_{net} = 0 = \text{Rate}_f - \text{Rate}_r$$

$$\text{Rate}_f = \text{Rate}_r$$

There are a few situations in which the net rate is equal to the forward reaction rate:

- One of the products escapes from the reaction system, so that its concentration remains at or very near zero at all times. A gaseous product that escapes from a solution is an example of this situation.
- The reaction has a large value of Keq, so that the reverse rate is negligible except near equilibrium, which is not achieved until almost all of the limiting reactant has been used.
- At the very beginning of a reaction in which only reactants are initially present. At time zero, the reverse rate must be zero because no products are present yet.

ONE-DIMENSIONAL NON-RELATIVISTIC CASE

Louis de Broglie made an analogy between matter waves and light waves, pointing out that wave packets of light change their velocity as the result of spatial variations in the index of refraction of the medium in which they are travelling. This behaviour comes about because the dispersion relation for light travelling through a medium with index of refraction n is $w = kc/n$, so that the group velocity, $u_g = dw/dk = c/n$. Thus, when n increases, u_g decreases, and vice versa.

The problem for matter waves is to determine how analogous modifications can be made to the free particle dispersion relation so as to produce accelerations of wave packets consistent with Newtonian dynamics in the geometrical optics limit. We make a simple guess as to how to modify the free particle dispersion relation for matter waves, limiting consideration initially to the one-dimensional, non-relativistic case. As in many situations in physics, the simple guess turns out to be correct! This leads to a connection between wave dynamics and Newtonian mechanics.

The dispersion relation for free matter waves is,

$$w = \left(k^2c^2 + \mu^2\right)^{1/2}$$

In the non-relativistic lim it $k^2c^2 \ll \mu^2$ and this equation can be approximated as,

$$w = \mu\left(1 + k^2c^2 / \mu^2\right)^{1/2} \approx \mu + k^2c^2 / (2\mu)$$

Let us modify this equation by adding a term *S'* which depends on *x*, and which plays a role analogous to the index of refraction for light:

$$k(x) = \left[\frac{2\mu\left[w - s(x)\right]}{c^2}\right]^{1/2}$$

The rest frequency has been made to disappear on the right side of the above equation by defining $S = S ¢ + \mathrm{m}$. This is done to simplify the notation.

Let us now imagine that all parts of the wave governed by this dispersion relation oscillate in phase. The only way this can happen is if *w* is constant, *i.e.*, it takes on the same value in all parts of the wave. It turns out we can do this if *S* is not a function of time. If *w* is constant, the only way *S* can vary with *x* in equation is if the wavenumber varies in a compensating way. Thus, constant frequency and spatially varying *S* together imply that $k = k(x)$. Solving equation for *k* yields

Since *w* is constant, the wavenumber becomes smaller and the wavelength larger as the wave moves into a region of increased *S*.

In the geometrical optics limit, we assume that *S* doesn't change much over one wavelength so that the wave remains reasonably sinusoidal in shape with approximately constant wavenumber over a few wavelengths. However, over distances of many wavelengths the wavenumber and amplitude of the wave are allowed to vary considerably. As yet we have no idea what causes *S* or where it comes from. For now we simply explore the consequences of its presence, especially in the geometric optics limit in which quantum dynamics gives way to Newtonian dynamics.

The group velocity calculated from the dispersion relation given by equation is,

$$\mu_g = \frac{dw}{dk} = \frac{kc^2}{\mu} = \left(\frac{2c^2(w - S)}{\mu}\right)^{1/2}$$

where *k* is eliminated in the last step with the help of equation. The resulting equation tells us how the group velocity varies as a matter wave traverses a region of slowly varying *S*. Thus, as *S* increases, u_g decreases and vice versa.

We can now calculate the acceleration of a wave packet resulting from the spatial variation in *S*. We assume that *x*(*t*) represents the position of the wave packet, so that $u_g = dx/dt$. Using the chain rule,

$$du_g / dt = \left(du_g / dx\right)\left(dx / dt\right) = \left(du_g / dx\right)^{\mu g}$$

we find,

$$a = \frac{du_g}{dt} = \frac{du_g}{dx}\mu_g = \frac{du_g^2 / 2}{dx} = \frac{c^2}{\mu}\frac{ds}{dx} = -\frac{\hbar}{m}\frac{ds}{dx}$$

The refraction of matter waves at a discontinuity in the potential energy. Let us suppose that the discontinuity occurs at $x = 0$. If the matter wave to the left of the discontinuity is,

$$\psi_1 = \sin\left(k_{1x}x + k_{1y}y - \omega_1 t\right)$$

and to the right is,

$$\psi_2 = \sin\left(k_{2x}x + k_{2y}y - \omega_2 t\right)$$

then the wavefronts of the waves will match across the discontinuity for all time only if $\omega_1 = \omega_2$ °w and $k_{1y} = k_{2y} \equiv k_y$. We are already familiar with the first condition from the one-dimensional problem, so the only new ingredient is the constancy of the component of the y wave vector. In two dimensions the momentum is a vector:

$$\Pi = mu \; when \left|u\right| \ll c,$$

where u is the particle velocity. Furthermore, the kinetic energy is

$$K = m\left|u\right|^2 / 2 = \left|\pi\right|^2 / (2m) = \left(\Pi_y^2\right) / (2m)$$

The relationship between kinetic, potential, and total energy is unchanged from the one-dimensional case, so we have

$$E = U + \left(\Pi_x^2 + \Pi_y^2\right) / (2m) = cons\tan t.$$

The de Broglie relationship tells us that $\Pi = \hbar k$, so the constancy of k_y across the discontinuity in U tells us that Π_y = constant. there.

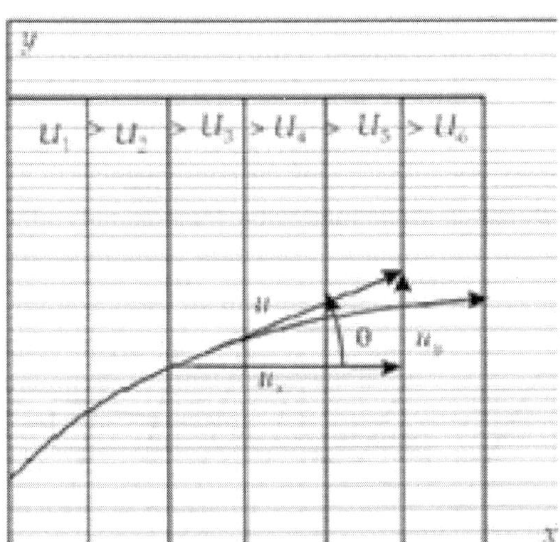

Fig. Trajectory of a wave packet through a variable potential energy, $U(x)$, which decreases to the right.

Let us now approximate a continuously variable $U(x)$ by a series of steps of constant U oriented normal to the x-axis. The above analysis can be applied at the jumps or discontinuities in U between steps, with the result that equations are valid across all discontinuities. If we now let the step width go to zero, these equations then become valid for U continuously variable in x. Using the procedure which we invoked before, we find the force components associated with $U(x)$ in the x and y directions to be $Fx = -dU/dx$ and $F_y = 0$. This generalizes to,

$$F = -\left(\frac{\partial U}{\partial x}, \frac{\partial U}{\partial y}, \frac{\partial U}{\partial z}\right)(3 - D\ conservative\ force)$$

in the three-dimensional case where the orientation of constant U surfaces is arbitrary. It is also valid when $U(x, y, z)$ is not limited to a simple ramp form, but takes on a completely arbitrary structure.

The definitions of work and power are slightly different in two and three dimensions. In particular, work is defined

$$W = F.\Delta x$$

where Δx is now a vector displacement. The vector character of this expression yields an additional possibility over the one dimensional case, where the work the force and the displacement of the object on which the force is acting are perpendicular to each other, the work done by the force is actually zero, even though the force and the displacement both have non-zero magnitudes. The power exhibits a similar change:

$$P = F.u.$$

Thus the power is zero if an object's velocity is normal to the force being exerted on it. Energy conservation by itself is somewhat less useful for solving problems in two and three dimensions than it is in one dimension. This is because knowing the kinetic energy at some point tells us only the magnitude of the velocity, not its direction.

If conservation of energy fails to give us the information we need, then we must revert to Newton's second law, as we did in the one-dimensional case. For instance, if an object of mass has initial velocity $u_0 = (u_0, 0)$ at location $(x, z) = (0, h)$ and has the gravitational potential energy $U = mgz$, then the force on the object is $F = (0, -mg)$. The acceleration is therefore $a = F/m = (0, -g)$. Since $a = du/dt = d2x/dt2$ where $x = (x, z)$ is the object's position, we find that

$$u = (C_1, -gt + C_2)\ x = (C_1 t + C_3, -gt_2/_2 + C_2 t + C_4),$$

where C_1, C_2, C_3, and C_4 are constants to be evaluated so that the solution reduces to the initial conditions at $t = 0$. The specified initial conditions tell us that $C_1 = u_0$, $C_2 = 0$, $C_3 = 0$, and $C_4 = h$ in this case. From these results we can infer the position and velocity of the object at any time.

❐

Index

Fermions 205
Ferromagnetism 154
Free Electron 160, 161, 162, 163